LAND SEARCH AND RESCUE ADDENDUM

LAND SEARCH AND RESCUE MANUAL

Volume I: Land Search and Rescue Addendum

National Search and Rescue Committee

Volume II: Catastrophic Incident Search and Rescue Addendum

National Search and Rescue Committee

Volume III: Incident Command System Field Operations Guide for Search and Rescue

Robert J. Koester

Volume IV: Management Field Operations Guide for Search and Rescue

Robert J. Koester

Volume V: Tactical Field Operations Guide for Search and Rescue

Robert J. Koester

LAND SEARCH AND RESCUE ADDENDUM

to the National Search and Rescue Supplement to the International Aeronautical and Maritime Search and Rescue Manual

Version 1.0

National Search and Rescue Committee

November 2011

Department of Homeland Security
Department of Interior
Department of Commerce
Department of Defense
Department of Transportation
National Aeronautics and Space Administration
Federal Communications Commission

dbS Productions
Charlottesville, Virginia

Cover photo: Photograph by Keith Lober used with written permission. Yosemite National Park Service Ranger Dov Bock shorthauls an injured climber from the East side of Mt. Ritter in the Sierra Nevada Range.

dbS

Published by dbS Productions LLC
P.O. Box 94
Charlottesville, Virginia USA 22902-0094
+1.434.293.5502 / 800.745.1581
www.dbs-sar.com

Library of Congress Cataloging-in-Publication Data

Land search and rescue addendum to the national search and rescue supplement to the International aeronautical and maritime search and rescue manual / National Search and Rescue Committee. -- Version 1.0.
p. cm. -- (Land search and rescue manual ; v. 1)
Includes index.
ISBN 978-1-879471-42-9 (entire series - bound : alk. paper) -- ISBN 978-1-879471-43-6 (perfect bound : alk. paper) -- ISBN 978-1-879471-44-3 (ebook) – ISBN 978-1-879471-53-5 (casebound)
1. Search and rescue operations--Handbooks, manuals, etc. I. National Search and Rescue Committee (U.S.) II. IAMSAR manual.

TL553.8.L36 2012
363.34'81--dc23

2012002192

Printed in the United States
10 9 8 7 6 5 4 3 2 1

Department of Defense
Department of Interior
Department of Commerce
Department of Homeland Security
Department of Transportation
Federal Communications Commission
National Aeronautics and Space Administration

National Search and Rescue Committee

Letter of Promulgation

**Land Search and Rescue Addendum (Version 1.0)
To the
National Search and Rescue Supplement**

Washington, D.C.
NOV 1 8 2011

This *Land SAR Addendum* was created by the National Search and Rescue Committee (NSARC) to provide standardized guidance and information on the search and rescue (SAR) of persons, vehicles, and missing aircraft operations. This Addendum further expands on the implementation of the *United States National Search and Rescue Plan (NSP),* the *National SAR Supplement (NSS) to the International Aeronautical and Maritime SAR (IAMSAR) Manual,* and is the land SAR equivalent to the *United States Coast Guard Search and Rescue Addendum to the NSS*, which provides detailed guidance for the conduct of SAR operations in the oceanic environment.

This is an important work in the standardization of land-based SAR operations. Information was painstakingly compiled from many sources and took into account decades of SAR best practices, as well as lessons learned from tragedies, mishaps, and successes in order to provide the best information available on the conduct of land SAR. For NSARC's member agencies, this Land SAR Addendum will be a foundational resource for identifying key agencies and organizations that support land-based SAR operations, how to plan searches, as well as the organization and implementation of successful lifesaving SAR operations.

This resource was developed for NSARC member agencies. However, from the beginning when the idea of a Land SAR Addendum was considered by NSARC, the member agencies had always remained committed to provide guidance that would benefit not only NSARC, but the States, Tribes, Territories and local SAR authorities who also save lives daily.

NSARC's mission has always been to improve interagency SAR coordination and cooperation, as well as the conduct of land-based SAR operations. It is with this purpose and goal that this Land SAR Addendum was created and freely given to all who are committed to lifesaving.

On behalf of the National Search and Rescue Committee,

Cari B. Thomas
Rear Admiral, United States Coast Guard
Director of Response Policy
Chair, National Search and Rescue Committee

This page intentionally left blank.

Foreword

Land SAR.

The National Search and Rescue Committee (NSARC) defines "land SAR" as the provisioning of civil SAR services for persons in distress or missing within the land areas of the U.S., including persons in distress in aircraft.

NSARC recognized there was a gap in national level guidance for land SAR planners and responders. As a result, NSARC agreed to develop a *Land Search and Rescue Addendum* to the *National SAR Supplement* (NSS) to the *International Aeronautical and Maritime Search and Rescue Manual* (IAMSAR Manual). This work was assigned to an NSARC Task Force led by the U.S. Air Force (USAF), National Park Service, National SAR School, and coordinated through the NSARC Secretariat.

The main objective of the Task Force was to draw on relevant best practices and established training and procedures to provide official guidance to Federal land SAR authorities. The Addendum was also envisioned to be used voluntarily for information, training, and operations within the non-Federal SAR community.

Although much of the information in this Addendum is unique to land SAR, the Addendum does not stand alone; it adds to relevant portions of the NSS and is a companion to other Addenda to the NSS.

After two years of extensive research and development of the principles, search theory, and practical guidance concerning land SAR operations, the NSARC Correspondence Work Group presents the Land SAR Addendum to all Federal SAR responders as foundational guidance for lifesaving operations.

Search theory, best practices shared, and lessons learned from previous SAR missions will continue to improve the SAR responder's ability to save lives. NSARC is committed to improving this Addendum when new information is obtained.

It is the NSARC Correspondence Work Group's desire that Federal SAR responders who conduct land SAR missions, as well as other State, Tribal, Territorial, and local SAR authorities, will find this Land SAR Addendum an invaluable resource in support of future lifesaving missions.

Correspondence Work Group
National Search and Rescue Committee

This page intentionally left blank.

Table of Contents

List of Figures

List of Tables

Glossary

aeronautical information publication (AIP): A publication issued by or with the authority of a State (nation) and containing aeronautical information of a lasting character essential to air navigation as defined by the International Civil Aviation Organization (ICAO). It is designed to be a manual containing thorough details of regulations, procedures and other information pertinent to flying aircraft in the particular country to which it relates. It is usually issued by or on behalf of the respective civil aviation administration.

aeronautical search and rescue: Search and rescue operations involving persons in distress aboard aircraft.

Aeronautical Information System – Replacement (AIS-R): A web-enabled, automated means for the collection and distribution of Service B messages, weather information, flight plan data, Notice to Airmen (NOTAM) messages, Pilot Report (PIREP) messages, and other operational information to all Federal Aviation Administration Air Traffic facilities. AIS-R is a direct replacement of the existing Aeronautical Information System (AIS) which has been in operation since 1999.

air route traffic control center (ARTCC): An FAA facility providing air traffic control service principally during en route flight to aircraft operating on IFR flight plans within controlled airspace.

air search area definition (ASAD): A method of determining the POC for a missing aircraft using statistical information from a 2009 NASA study of crash locations.

air traffic control (ATC): A service operated by an appropriate authority to promote the safe, orderly, and expeditious flow of air traffic.

aircraft coordinator (ACO): A person who coordinates the involvement of multiple aircraft in SAR operations.

alert notice (ALNOT): A message sent by an FSS or an ARTCC that requests an extensive communications search for overdue, unreported, or missing aircraft. This is an FAA term.

Alert Phase: A situation wherein *apprehension* exists as to the safety of an aircraft or marine vessel, and of the persons on board.

alert: Report of an apparent distress routed to the search and rescue system.

ambiguity resolution: In the Cospas-Sarsat system, the process of determining which of the available computed Doppler positions and/or encoded positions for a transmitting distress beacon is the resolved (confirmed or composite) position.

automatic direction finding (ADF): Equipment that determines bearing to a radio station;

average maximum detection range (AMDR): Is the distance on average that a sensor (searcher) has the maximum detection range to detect (see) an object in specific environmental conditions as well as vegetation or any ground cover or other obstacles in their search segment.

awareness stage: A period during which the SAR system becomes aware of an actual or potential incident.

beacon ID (Cospas-Sarsat): The 15 hexadecimal characters that uniquely identify each 406 MHz distress beacon.

catastrophic incident: Any natural or manmade incident, including terrorism, that results in extraordinary levels of mass casualties, damage, or disruption severely affecting the population, infrastructure, environment, economy, national morale, and/or government functions.

catastrophic incident search and rescue (CISAR): Civil SAR operations carried out as all or part of the response to an emergency or disaster declared by the President, under provisions of the NRF and ESF #9 Annex

civil search and rescue (civil SAR): Search operations, rescue operations, and associated civilian services provided to assist persons and property in potential or actual distress in a non-hostile environment.

command center (CC): Multi-mission center that may perform the function of a rescue coordination center in addition to having staff and capabilities to perform other functions.

commence search point (CSP): Point normally specified by the SMC where a SAR facility is to begin its search pattern.

community emergency response team (CERT): A CERT is a group of people organized as a neighborhood-based team that receives special training to enhance their ability to recognize, respond to, and recover from a major emergency or disaster situation. Teams are trained by professional responders in areas that will help them take care of themselves and others before, during and after a major emergency. As an organized team, individuals can provide vital services in the absence of and while waiting for the arrival of emergency responders, and they often assist once responders arrive.

composite position: In the Cospas-Sarsat system, a position generated by the USMCC by combining data from two or more satellite passes over a beacon. Any ambiguity from previous satellite passes is resolved and solutions from different satellite passes merged to refine the position of the beacon.

conclusion stage: A period during a SAR incident when SAR facilities return to their regular location and prepare for another mission. Conclusion stage is also known as Demobilization Phase.

conditional probability: The probability of an event given another event does in fact exist or has in fact occurred. (Example: POD is the conditional probability that a sensor will detect a target given the target is in area being searched.)

confinement: A search strategy (with many possible tactics) to limit the possible area into which the search objective might travel undetected, or to establish that the search objective has not already passed a desired search area boundary.

consensus: a method to bring together and consider the experiences, judgments, and opinions of the search experts on the leadership team in order to develop a common ground on various parts of a search plan; including, scenarios and their relative weights, the search area, regions, and probability of containment.

Cospas-Sarsat system: A satellite system designed to detect distress beacons transmitting on the 406 MHz frequency.

course: The intended horizontal direction of travel.

coverage factor (C): The ratio of the search effort (Z) to the area searched (A). C = Z/A. For parallel sweep searches, it may be computed as the ratio of sweep width (W) to track spacing (S). C = W/S.

critical incident stress: Any event which overwhelms the capacities of a person to psychologically cope with an incident; the normal physical and psychological reactions of normal people experiencing an abnormal event or demand.

Critical Incident Stress Management (CISM): A comprehensive, integrated, systematic, multi tactic form of crisis intervention that is applied to manage critical incident stress after traumatic events.

critique: A thorough review of an operation, both positive and negative, in order to determine the effectiveness and weaknesses of the preplan; used to assist in the development of action items to improve the preplan, as well as determine future training, equipment requirements, and processes for the next incident.

cumulative probability of success (POS_{cum}): The accumulated probability of finding the search object with all the search effort expended over all searches to date. POS_C is the sum of all individual search POS values.

datum: A geographic point, line, or area used as a reference in search planning or mapping.

Direct User Access Terminal (DUAT): **DUATS** is a weather information and flight plan processing service contracted by the Federal Aviation Administration (FAA) for use by U.S. civil pilots and other authorized users. The DUAT Service is a telephone- and Internet-based system which allows the pilot to use a personal computer for access to a FAA database to obtain weather and aeronautical information and to file, amend, and cancel domestic IFR and VFR flight plans; and close domestic VFR flight plans. The DUAT Service provides direct access to weather information via a National Airspace System (NAS) Data Interchange Network II (NADIN-II) interface to the Weather Message Switching Center Replacement (WMSCR) System and the Air traffic Control (ATC) Facilities for filing flight plans.

direction finding (DF): radio receiver that indicates the direction from which a signal is coming.

distress beacon: Device operating on 121.5 MHz, 243 MHz, or 406 MHz intended solely for distress signaling.

Distress Phase: A situation wherein there is reasonable certainty that a vessel or other craft, including an aircraft or a person, is threatened by grave and imminent danger and requires immediate assistance.

Domestic Event Network (DEN): A 24/7 FAA sponsored, telephonic conference call network that includes all of the Air Route Traffic Control Centers (ARTCC) in the U.S. It also includes various other governmental agencies that monitor the DEN. The purpose of the DEN is to provide timely notification to the appropriate authorities that there is an emerging air-related problem or incident within the continental U.S.

Doppler Shift: Change in the apparent frequency of a wave as observer and source move toward or away from each other. Doppler Shift is used by the Cospas-Sarsat system to determine the position of a distress beacon that has been energized.

effective sweep width (ESW): The area under the lateral range curve. It is twice the range of a definite range law detection, such that the sensor and an equivalent definite range law sensor detect on the average the same number of uniformly distributed targets.

elemental solution (Cospas-Sarsat): Single satellite pass solution, also used to determine a composite position.

emergency locator transmitter (ELT): Aeronautical radio distress beacon for alerting and transmitting homing signals.

emergency phase: A generic term, that describes the uncertainty phase, alert phase, or distress phase of a potential or ongoing SAR mission.

emergency position-indicating radio beacon (EPIRB): A device, usually carried aboard maritime craft, that transmits a signal that alerts search and rescue authorities and enables rescue units to locate the scene of the distress.

emergency support function (ESF): Provides the structure for coordinating Federal interagency support for a Federal response to an incident. They are mechanisms for grouping functions most frequently used to provide Federal support to States and Federal-to-Federal support, both for declared disasters and emergencies under the Stafford Act and for non-Stafford Act incidents.

extended communication search (EXCOM): Comprehensive communications search to find information or clues about the location of missing persons. Normally conducted after a PRECOM has yielded no results, or when the mission is upgraded to the Alert phase.

false alert: Distress alert received from any source, including communications equipment intended for alerting, when no distress situation actually exists, and a notification of distress should not have resulted.

first alert: In the Cospas-Sarsat system, the first detection of a distress beacon by the Cospas-Sarsat System, position information may or may not be available.

flight plan: Specified information, relating to the intended flight of an aircraft, that is filed orally or in writing with air traffic control.

flight service station (FSS): An air traffic facility that provides en route communications and VFR SAR services, assists lost aircraft and aircraft in emergency situations, and originates Notices to Airmen.

forward-looking infrared radar (FLIR): An imaging system, mounted on board surface vessels or aircraft, designed to detect thermal energy (heat) emitted by search objects and convert it into a visual display. (IAMSAR Manual)

geostationary Earth orbiting local user terminal (GEOLUT): In the Cospas-Sarsat system, a ground receiving station that detects, processes, and recovers the coded transmissions of 406 MHz distress beacons, and relays the appropriate information to a Mission Control Center

Global Positioning System (GPS): A specific satellite-based system used in conjunction with mobile equipment to determine the precise position of the mobile equipment.

grid cell: A square or rectangular area formed by pairs of adjacent, perpendicular, grid lines.

grid: Any set of intersecting perpendicular lines spaced at regular intervals.

ground speed (GS): The speed an aircraft is making relative to the earth's surface.

hasty search: see 'rapid search'

heading: The horizontal direction in which a craft is pointed.

homeland defense (HD): An activity undertaken for the military protection of the territory or domestic population of the United States, or of infrastructure or other assets of the United States determined by the Secretary of Defense as being critical to national security, from a threat or aggression against the United States (32 USC 901).

image (mirror image): In the Cospas-Sarsat system, of the two positions associated with a single satellite pass, the one from which a signal is not emanating.

incident: A SAR case involving investigation and telephone communications searches; generally associated with the Uncertainty Phase of a search.

incident action plan (IAP): Contains objectives reflecting the overall incident strategy and specific tactical actions and supporting information for the next operational period. The plan may be oral or written. When written, the plan may have a number of forms as attachments (e.g., traffic plan, safety plan, communications plan, map, etc.). See also **search action plan**.

incident command system (ICS): (1) An all-risk on-scene system of coordination for any type of incident involving multiple responding authorities and multiple jurisdictions.(2) A standardized on-scene emergency management concept specifically designed to allow its user(s) to adopt an integrated organizational structure equal to the complexity and demands of single or multiple incidents, without being hindered by jurisdictional boundaries.

incident commander (IC): The individual responsible for the management of all incident operations at the incident site. See also **on scene coordinator (OSC)**.

information request (INREQ): A message request for information about an unreported or overdue aircraft in United States domestic airspace. Corresponds to the declaration of the Uncertainty phase. This is an FAA term.

initial action stage: A period during which preliminary action is taken to alert SAR facilities and obtain amplifying information.

initial reflex search: See **rapid search**.

instrument flight rules (IFR): Rules governing the procedures for conducting instrument flight. Also a term used by pilots and controllers to indicate type of flight plan. The term "IFR" is also used in the U.S. to indicate weather conditions less than minimum VFR requirements.

instrument meteorological conditions (IMC): Meteorological conditions expressed in terms of visibility, distance from clouds, and ceiling less than minimum VFR requirements.

initial planning point (IPP): The point that is initially used to plan the search incident. The IPP may be the original Point Last Seen (PLS) or the Last Known Position (LKP).The IPP may also be a point entirely separate based on the best available investigative information. IPP does not move after planning begins.

investigation: The systematic inquiry and examination of all aspects and information surrounding a SAR case; includes all activities outside of the active field search for clues and for the search object.

knot (kt): A unit of speed equal to one nautical mile per hour.

last known position (LKP): (1) It is the last substantiated position based on clues or evidence belonging to the missing subject and indicating that the missing subject was known to be at that location. (2) Last witnessed, reported, or computed position of a distress person or craft.

lateral range curve: Graph of the probability of detection as a function of lateral range.

local user terminal (LUT): An earth receiving station that receives beacon signals relayed by Cospas-Sarsat satellites, processes them to determine the location of the beacons, and forwards the signals.

location: Exact geographical position of a distress beacon or other search object on the surface of the earth.

low Earth orbit local user terminal (LEOLUT): A ground receiving station in the Cospas-Sarsat system that detects, characterizes and locates distress beacons, and forwards the appropriate information to an MCC.

mass rescue operations (MRO): Civil search and rescue services characterized by the need for immediate response to large numbers of persons in distress, such that the capabilities normally available to search and rescue authorities are inadequate.

merge: In the Cospas-Sarsat system, the process of developing composite solutions by combining matched single pass (elemental) solutions.

meteorological visibility: The maximum range at which a large object, such as landmasses or mountains, can be seen. Also referred to as meteorological range.

mission: A SAR case in which facilities have been committed to actively search; generally associated with the Distress Phase of a search.

mission control center (MCC): In the Cospas-Sarsat system, the station that accepts alert messages from the local user terminal(s) and other mission control centers to distribute to the appropriate rescue coordination centers or other search and rescue points of contact.

National Incident Management System (NIMS): A system prescribed by Homeland Security Presidential Directive 5 to coordinate emergency preparedness and incident management among various federal, state, and local agencies. NIMS provides a uniform nationwide approach to work effectively and efficiently together to prepare for, respond to, and recover from domestic incidents no matter what the cause, size or complexity, including catastrophic acts of terrorism and disasters.

National Response Framework (NRF): A guide to how the Nation conducts all-hazards response. It is built upon scalable, flexible, and adaptable coordinating structures to align key roles and responsibilities across the Nation, linking all levels of government, nongovernmental organizations, and the private sector. It is intended to capture specific authorities and best practices for managing incidents that range from the serious but purely local, to large-scale terrorist attacks or catastrophic natural disasters.

National Search and Rescue Committee (NSARC): Federal committee comprised of the Departments of Homeland Security, Defense, Transportation, Interior, and Commerce, Federal Communications Commission, and the National Aeronautics and Space Administration. Established to oversee the National Search and Rescue Plan and act as a coordinating forum for national SAR matters.

National Search and Rescue Plan (NSP): An interagency agreement providing national arrangements for coordination of search and rescue services to meet domestic needs and international commitments.

National Special Security Event (NSSE): A designated event that, by virtue of its political, economic, social, or religious significance, may be the target of terrorism or other criminal activity.

National Track Analysis Program (NTAP): An FAA system for retrieval of computer-stored radar data to locate a missing aircraft's last position.

National Wildfire Coordinating Group (NWCG): A group formed under the direction of the Secretaries of the Interior and Agriculture to improve the coordination and effectiveness of wildland fire activities and provide a forum to discuss, recommend appropriate action, or resolve issues and problems of substantive nature.

on scene coordinator (OSC): A person designated to coordinate search and rescue operations within a specified search area. See also **incident commander (IC)**.

on scene endurance: The amount of time a search unit may spend at the scene engaged in SAR activities.

on scene: The search area or the actual distress site.

operational period (ICS): The period of time scheduled for execution of a given set of operational objectives as specified in the Incident Action Plan. Operational periods can be of various lengths, although usually not over 24 hours.

operations stage: A period during a SAR incident when SAR facilities proceed to the scene, conduct search, rescue survivors, assist distressed craft, provide emergency care for survivors, and deliver survivors to a suitable facility.

optimal search area: The search area which will produce the highest probability of success when searched uniformly with the search effort available.

optimal search plan: A plan that maximizes the probability of success of finding the search object using the available search effort.

overdue: A situation where a search object has failed to arrive at its intended destination when expected and remains missing.

overlap zone: In the Cospas-Sarsat system, an area of predetermined width common to two or more service areas.

personal locator beacon (PLB): Personal radio distress beacon used by individuals for distress alerting.

place of safety: Location where rescue operations are considered to terminate and where: 1) The survivor's safety or life is no longer threatened; 2) basic human needs (such as food, shelter and medical needs) can be met; and (3) transportation arrangements can be made for the survivor's next or final destination.

planning stage: A period during a SAR incident when an effective plan of operations is developed.

point last seen (PLS): It is the point where the search object was last seen by a witness, or captured on video at a specific time and location.

position (location): A geographical location normally expressed in degrees and minutes of latitude and longitude (or USNG Coordinates).

posse comitatus: The Posse Comitatus Act (18 U.S.C. 1385) prohibits the use of any part of the Army or the Air Force to execute or enforce the laws, except as authorized by the Constitution or Act of Congress; includes prohibition of direct participation in a search, seizure, arrest, or other similar activity. This restriction is in addition to the constitutional limitations of the power of the Federal Government at the local level.

possibility area: (1) The smallest area containing all possible survivor or search object locations. (2) For a scenario, the possibility area is the smallest area containing all possible survivor or search object locations that are consistent with the facts and assumptions used to form the scenario.

preliminary communication search (PRECOM): Initial limited communications check, normally directed by the SMC during the Uncertainty phase, of areas where the missing persons may be.

pre-plan: Guidelines and current information established before a search that allow a quick, effective response, allow strategic decisions to be made outside the heat of the moment, and allow flexibility within a well thought-out boundary of reasonable actions. Decisions specific to a particular search, including tactical operations, should be contained in a search action plan or incident action plan, as opposed to a search pre-plan.

probability along track (P_T): The likelihood the missing craft is located between the last known position and the given percentage of the distance of the intended or expected route of flight; based on ASAD statistical data and assumes the entire length of the route of flight from last known position to intended destination, not just the portion being searched.

probability density (pDen): Probability of containment per unit area; often POC/mi^2. pDen is constant throughout any given region.

probability map: A set of grid cells covering a scenario's possibility area where each grid cell is labeled with the probability of the search object being in that grid cell. That is, each grid cell is labeled with its own POC value.

probability of area (POA): See probability of containment (POC). *(Note: POA and POC are synonymous.)*

probability of containment (POC): The probability that the search object is contained within the boundaries of an area, sub-area, or grid cell. *(Note: POA and POC are synonymous.)*

probability of detection (POD): The probability of the search object being detected, assuming it was in the areas that were searched. POD is a function of coverage factor, sensor, search conditions and the accuracy with which the search unit navigates its assigned search pattern. Measures sensor effectiveness under the prevailing search conditions.

probability of offset (P_o): The likelihood the missing aircraft is located within the given offset distance from the centerline of the intended or expected track or route of flight; based on ASAD statistical data and assumes equal likelihood left and right of track centerline.

probability of success (POS): The probability of finding the search object with a particular search. For each sub-area searched, POS = POC x POD. Measures search effectiveness.

ramp check: a telephone call requesting the responsible agency at an airport physically check the field and hangars for a specific aircraft being sought.

rapid search: Also known as 'hasty search' and 'initial reflex search.' Quickly looking for the search object along likely routes of travel, linear geographical features or in high probability locations.

RCC controller: The SAR mission coordinator's duty officer in the RCC.

region POC: The probability of containment of an individual region; often determined by a consensus of the on-scene leadership team.

region: A delineated portion or subset of the search area within which there is a nearly uniform distribution of likelihood for containing the missing object; that is, the missing object is equally likely to be at any point in a given region, although different regions may have different likelihoods of containment. Thus, probability density (pDen) is constant within any given region. Regions and region boundaries are based solely on the factors that affect the location and POC of the missing object; factors affecting searchers are considered in defining segments, not regions. Often, regions are further subdivided into segments in order to be effectively searched. Region boundaries should be distinguishable by searchers (e.g., stream, drainage, ridge, natural geographical feature, etc.).

rescue: An operation to retrieve persons in distress, provide for their initial medical or other needs, and deliver them to a place of safety.

rescue coordination center (RCC): A unit responsible for promoting efficient organization of SAR services and for coordinating the conduct of SAR operations within a search and rescue region.

rescue sub-center (RSC): A unit subordinate to an RCC established to complement the latter according to particular provisions of the responsible authorities.

responsible authority: The government agency or agencies who have legal responsibility for finding missing persons and have jurisdiction over the area where the person becomes missing.

SAR Coordinator (SC): One or more persons or agencies within an Administration with overall responsibility for establishing and providing SAR services, and ensuring that planning for those services is properly coordinated.

SAR facility: (1) Any mobile resource, including designated SRUs, used to conduct SAR operations. (2) Fixed facilities for the incident; may include the Incident Base, feeding areas, sleeping areas, sanitary facilities, etc.

SAR incident: Any situation requiring notification and alerting of the SAR system and which may require SAR operations.

SAR Mission Coordinator (SMC): The official temporarily assigned to coordinate response to an actual or apparent distress situation.

SAR mission: Any SAR situation involving dispatch of SAR resources.

SAR plan: A general term used to describe documents which exist at all levels of the national and international SAR structure to describe goals, arrangements, and procedures which support the provision of SAR services. See also **search action plan** and **incident action plan**.

SAR stage: Typical steps in the orderly progression of SAR missions. These are normally Awareness, Initial Action, Planning, Operations, and Mission Conclusion.

scenario: a consistent set of known facts and assumptions describing what may have happened to the survivors.

search: An operation, normally coordinated by an RCC or RSC, using available personnel and facilities to locate persons in distress.

search action plan: Message, normally developed by the SMC, for passing instructions to SAR facilities and agencies participating in a SAR mission. See also **incident action plan**.

search and rescue region (SRR): An area of defined dimensions, associated with an RCC, within which SAR services are provided.

search and rescue unit (SRU): A unit composed of trained personnel and provided with equipment suitable for the expeditious conduct of SAR operations.

search area: The area, determined by the search planner, that is to be searched. This area may be sub-divided into search sub-areas for the purpose of assigning specific responsibilities to the available search facilities.

search effort (Z): A measure of the [effective] area a search facility can search within the limits of search speed, endurance, and sweep width. Search effort is computed as the product of search speed (V), search endurance (T), and sweep width (W).
Z = (V) X (T) X (W)

search endurance (T): The amount of "productive" search time available at the scene. This figure is usually taken to be 85% of the on scene endurance, leaving a 15% allowance for investigating sightings and navigating turns at the ends of search legs.

search object: A ship, aircraft, person or other craft missing or in distress or survivors or related search objects or evidence for which a search is being conducted.

search pattern: A trackline or procedure assigned to an SRU for searching a specified area.

search speed (V): The speed (or velocity) with which a search facility moves over the ground when searching.

search sub-area: A designated area to be searched by a specific assigned search facility or possibly two facilities working together in close coordination.

search subject: The person being searched for, commonly referred to as the "Subject". Searches focusing primarily on the discovery of objects as opposed to person, e.g. aircraft or vessels, commonly utilize the term "search object".

segment: (1) A delineated portion or subset of the search area bounded and sized to be searchable by assigned facilities. Segments and their boundaries are based on factors that affect the searchers. For the purposes of tracking POC and search progress, it is generally inadvisable to allow a segment to cross-region boundaries. See also **region**. (2) In ICS, a segment is a geographical area in which a task force/strike team leader or supervisor of a single resource is assigned authority and responsibility for the coordination of resources and implementation of planned tactics. A segment may be a portion of a division or an area inside or outside the perimeter of an incident.

segment POC: The probability of containment of an individual segment; determined as a proportion of its region POC, based on the ratio of the area of the segment to the area of the region. Segment POC cannot be reliably established if a segment crosses region boundaries.

sensors: Human senses (sight, hearing, touch, etc.), those of specially trained animals (such as dogs), or electronic devices used to detect the object of a search.

sortie: Individual movement of a resource in conducting a search or rendering assistance.

sound search: Synonymous with Sound Sweep, with the latter being the proper term for the specific grid-type technique. The use of sound (e.g. calling a subject's name), in conjunction with other types of direct searching tactics is common, but should not be referred to as a "sound search".

sound sweep: An effective search technique that combines a sound attraction tactic with the direct tactic of wide spaced grid searching. The use of sound, typically whistle blasts, is closely choreographed with simultaneous blasts followed by a uniform listening period for subject response. The use of this technique presumes a responsive subject and, thus, is most effectively utilized in the early stages of a search.

strategy: The general plan or direction selected to accomplish incident objectives.

sweep width (W): A measure of the effectiveness with which a particular sensor can detect a particular object under specific environmental conditions.

time of closest approach (TCA): Time during a satellite pass when the satellite is closest to a signal source.

track spacing (S): The distance between adjacent parallel search tracks.

true air speed (TAS): The speed an aircraft is traveling through the air mass. TAS corrected for wind equals ground speed.

Uncertainty Phase: A situation wherein *doubt* exists about the safety of an aircraft or a marine vessel, and of the persons on board.

unified command: In ICS, Unified Command is a unified team effort which allows all agencies with responsibility for the incident, either geographical or functional, to manage an incident by establishing a common set of incident objectives and strategies. This is accomplished without losing or abdicating agency authority, responsibility, or accountability.

vector: A graphic representation of a physical quantity or measurement, such as wind velocity, having both magnitude and direction.

visual flight rules (VFR): Rules governing procedures for conducting flight under visual meteorological conditions. In addition, used by pilots and controllers to indicate type of flight plan. (The term "VFR" is also used in the U. S. to indicate weather conditions equal to or greater than minimum VFR requirements.)

visual meteorological conditions (VMC): Meteorological conditions expressed in terms of visibility, distance from cloud, and ceiling meeting or exceeding the minimums specified for VFR requirements.

List of Acronyms

A	search area (can be subscripted for regions and segments)
A/C	aircraft
ADF	automatic direction finding
ADS	Air Defense Sector
AFRCC	Air Force Rescue Coordination Center
AIP	aeronautical information publication
AIS-R	Aeronautical Information System-Replacement
AKRCC	Alaska Rescue Coordination Center
ALNOT	alert notice
AMDR	average maximum detection range
ANG	Air National Guard
AOPA	Aircraft Owners and Pilots Association
ARCC	aeronautical RCC
ARNG	Army National Guard
ARTCC	air route traffic control center
ASAD	air search area definition
ATC	Air Traffic Control
ATCT	Airport Traffic Control Tower
ATO	Air Traffic Organization
ATS	air traffic service
ATV	all terrain vehicle
AWLD	Aircraft Wreckage Locator Database
C	coverage factor
CAP	Civil Air Patrol
CB	citizens band
CDRNORTHCOM	Commander, U.S. Northern Command
CDRUSPACOM	Commander, U.S. Pacific Command
CERT	community emergency response team
CFR	Code of Federal Regulations
CGAUX	Coast Guard Auxiliary
CIS	Critical Incident Stress
CISAR	catastrophic incident search and rescue
CISM	critical incident stress management
CONUS	continental United States
Cospas	Cosmicheskaya Sistyema Poiska Avariynych Sudov (Russian for "Space System for Search of Vessels in Distress")
CSP	commence search point
DCO	Defense Coordinating Officer
DEN	Domestic Event Network
DF	direction finding
DMB	datum marker buoy
DOC	Department of Commerce
DOD	Department of Defense
DOI	Department of Interior
DOT	Department of Transportation
DSCA	defense support of civil authorities
DUAT	Direct User Access System
EADS	Eastern Air Defense Sector
EICC	Emergency Incident Coordination Center
ELT	emergency locator transmitter
EMAC	Emergency Management Assistance Compact
EMS	emergency medical services
EMT	emergency medical technician
EOC	emergency operations center

EPIRB	emergency position-indicating radio beacon
EPLO	Emergency Preparedness Liaison Officer
ESF	emergency support function
ESW	effective sweep width
ETA	estimated time of arrival
ETD	estimated time of departure
EXCOM	extended communication search
FAA	Federal Aviation Administration
FAR	federal aviation regulation
FBI	Federal Bureau of Investigation
FCC	Federal Communications Commission
FDOA	frequency difference of arrival
FEMA	Federal Emergency Management Agency
FLIP	flight information publication
FLIR	forward-looking infrared
FM	frequency modulation
FSH	flight service hub
FSS	flight service station
GA	general aviation
GARS	global area reference system
GEO	geostationary Earth orbit
GEOLUT	geostationary earth orbit local user terminal
GMT	Greenwich Mean Time
GNSS	global navigation satellite system
GOES	geo-synchronous orbiting earth satellite
GPS	global positioning system
HD	Homeland Defense
HSPD	Homeland Security Presidential Directive
IAMSAR Manual	International Aeronautical and Maritime Search and Rescue Manual
IC	Incident Commander
ICAO	International Civil Aviation Organization
ICP	incident command post
ICS	incident command system
IDEA	integrated detection experiment assistant
IFF	identification, friend or foe
IFR	instrument flight rules
IMO	International Maritime Organization
INREQ	information request
IPP	initial planning point
IST	incident support team
JRCC	joint rescue coordination center
KHz	kilohertz
km	kilometers
kt	knot (nautical miles per hour)
L	length
l	search sub-area length
LEO	low Earth orbit
LEOLUT	low Earth orbit local user terminal
LKP	last known position
LOS	line of sight
LUT	local user terminal
m	meters
MCC	mission control center
MEO	medium Earth orbit
MGRS	Military Grid Reference System
MHz	megahertz
MOA	memorandum of agreement
MOU	memorandum of understanding
MRA	Mountain Rescue Association
MRCC	maritime rescue coordination center
MRO	mass rescue operation
n	number of required track spacings
N	number of SAR resources
NASA	National Aeronautics and Space Administration
NASAR	National Association for Search and Rescue

NAWAS	National Alert Warning System
NIFOG	National Interoperability Field Operations Guide
NIMS	National Incident Management System
NIUSR	National Institute for Urban Search and Rescue
nm	nautical mile
NOAA	National Oceanic and Atmospheric Administration
NOTAM	notice to airmen
NPS	National Park Service
NRF	National Response Framework
NSARC	National Search and Rescue Committee
NSP	National Search and Rescue Plan
NSS	National Search and Rescue Supplement
NSSE	National Special Security Event
NTA	National Track Analysis
NTAP	National Track Analysis Program
NTSB	National Transportation Safety Board
NVOAD	National Voluntary Organizations Active in Disaster
OPOS	overall probability of success
$OPOS_{cum}$	cumulative overall probability of success
OSC	on scene coordinator
OSE	on scene endurance
P	parallel pattern
pDen	probability density
PIO	public information officer
PLB	personal locator beacon
PLS	point last seen
POA	probability of area (replaced by POC)
POB	persons on board
POC	probability of containment; can be subscripted for regions, segments, and search periods
POD	probability of detection can be subscripted for regions and segments
POD_{cum}	cumulative probability of detection
POS	probability of success (can be subscripted for regions and segments)
POS_{cum}	cumulative probability of success
PR	personnel recovery
PRECOM	preliminary communication search
PSR	probable success rate
QALQ	Data Request On Aircraft
RADES	radar evaluation squadron
RANPs	regional air navigation plans
RAP	rescue action plan
RCC	rescue coordination center
RDF	radio direction finder
RF	radio frequency
RHIB	rigid hull inflatable boat
RSC	rescue sub-center
SAP	search action plan
SAR	search and rescue
SARSAT	Search and Rescue Satellite-Aided Tracking
SATCOM	satellite communications
SC	SAR coordinator
SDSA	subjective deductive search area
SEND	satellite emergency notification device
SITREP	situation report
SMC	SAR mission coordinator
SPOC	search and rescue point of contact
SRR	search and rescue region
SRS	search and rescue sub-region
SRU	search and rescue unit
SSA	statistical search area
T	search time available

TACAN	Tactical Air Navigation
TDOA	time difference of arrival
TFR	temporary flight restriction
TRACON	Terminal Radar Approach Control
TSO	technical standard order
UC	unified command
UHF	ultra high frequency
USAF	United States Air Force
USAR	urban search and rescue (may also be written as US&R)
USC	United States Code
USCG	United States Coast Guard
USMCC	United States Mission Control Center
USN	United States Navy
USNG	United States National Grid
UTC	coordinated universal time
UTM	universal transverse Mercator grid
UTV	utility vehicle
V	SAR facility ground speed
VFR	visual flight rules
VHF	very high frequency
VMC	visual meteorological conditions
w	search sub-area width
W	sweep width
WADS	Western Air Defense Sector
Z	effort; effective area swept
Zt	total available effort

Introduction

The *National Search and Rescue Plan* (NSP), signed by the member agencies of the National Search and Rescue Committee (NSARC), was originally signed and implemented over 50 years ago. Even with the passing of time, the purpose, responsibilities, and mission of the signatory agencies has always remained the same: a commitment to always improve the interagency coordination and conduct of search and rescue (SAR) operations in the United States and globally.

The primary implementing guidance for the NSP is provided in the *International Aeronautical and Maritime Search and Rescue Manual (IAMSAR) Manual* and the United States *National Search and Rescue Supplement* (NSS) to the IAMSAR Manual.

This *Land Search and Rescue Addendum*, to be used in conjunction with both the IAMSAR Manual and the NSS, has been also developed as a foundational resource for SAR managers, planners, and the lifesaver. NSARC's purpose in developing the Land SAR Addendum was to provide standardized guidance and information concerning land-based SAR operations.

The two general categories of land SAR discussed in this Addendum are:

- Aeronautical SAR involving persons in distress in aircraft on or over land areas, particularly in wilderness areas; and
- Search and rescue of persons missing or in distress.

The guidance within this Addendum is intended to assist SAR personnel without preempting their experience and sound judgment. SAR planning in particular is both an art and a science, relying greatly on the creativity and experience of the personnel involved. Because of the many variables encountered during SAR operations and the uniqueness of each individual SAR case, any guidance must be tempered with sound judgment, having regard for the individual situation. Nothing in this Addendum should be construed as relieving SAR personnel of the need for initiative and sound judgment. Therefore, few actions or procedures discussed in this Addendum are mandatory.

Of importance to the SAR planner is the standardization of search planning theory. Extensive work has been conducted in this field to provide guidance and information that explains the procedures for determining, based on the available SAR resources, how to plan for and implement a search. Training on search theory and search planning is available from the National SAR School (Inland SAR Planning course) and the Air Force Rescue Coordination Center (Basic Inland SAR course).

This page intentionally left blank.

Part 1: U.S. SAR System

This page intentionally left blank.

Section 1-1: Land SAR

Overview and General Concepts

The provisions of this Addendum are considered guidance for Federal civil Search and Rescue (SAR) responders of the National SAR Committee (NSARC) in the conduct of land SAR operations. Although there is no obligation for State, Tribal, Territorial or local authorities to follow this guidance, NSARC invites all SAR responders to use this Addendum as baseline guidance in order to support land-based lifesaving operations.

The *National Search and Rescue Supplement (NSS)*, its associated Addendums, and the *International Aeronautical and Maritime Search and Rescue (IAMSAR) Manual,* will also expand on the general information in this Section.

SAR services include the performance of distress monitoring, communication, coordination, as well as the provisioning of medical advice, initial medical assistance, and/or medical evacuation, through the use of public and private resources, including cooperating aircraft and SAR teams.

SAR is comprised of search functions and rescue functions. This is an important point not only because of differences in skills and resources involved between these two types of operations, but sometimes due to differences in responsibilities, jurisdictions, and legal aspects. *Search* is an operation that uses available personnel and facilities to locate persons in distress. *Rescue* is an operation to retrieve persons in distress, provide for their initial medical or other needs, and deliver them to a place of safety.

Civil SAR operations, including emergency medical aid, are conducted in a wide variety of environments. These operations often require extended response times and use of specialized teams and equipment.

Cross jurisdiction rescues should normally be allowed when:

- The responsible jurisdiction is not able to immediately respond;
- The location of the distress is reasonably well-known; and
- A suitable rescue resource is available to conduct the rescue.

Land SAR, as discussed in this Addendum, is the provision of civil SAR services for persons in distress or missing within the land areas of the United States, including aircraft. These searches involve locating persons in distress and are normally conducted by State or local authorities. If additional assistance is required, the appropriate Federal *Rescue Coordination Center (RCC)* or the *National Park Service (NPS)* can coordinate use of available Federal SAR facilities.

SAR Coordinator (SC)

Federal SAR responsibilities under the *National Search and Rescue Plan (NSP)* are assigned to *SAR Coordinators (SC)*. The SC is a person or agency with overall responsibility for establishing and providing SAR services for one or more *Search and Rescue Regions (SRRs)* for which the U.S. has primary responsibility. Table 1-1-1 details the assignment of Federal SCs in the NSP and other SAR responsibilities.

Table 1-1-1: U.S. Federal SC Responsibilities **(Reference: National SAR Plan)**	
U.S. Northern Command (CDRUSNORTHCOM)	Recognized SC for the U.S. aeronautical SRR corresponding to the continental U.S. other than Alaska. Commander, U.S. Northern Command has delegated the routine Federal aeronautical SAR Coordination authority for the Langley SRR to Commander, Air Forces North/Joint Force Air Component Commander.
U.S. Pacific Command (CDRUSPACOM)	Recognized SC for the U.S. aeronautical SRR corresponding to Alaska. SAR coordinator responsibility for Alaska has been delegated from CDRUSPACOM to CDRPACAF to Commander of 11th Air Force. 11 AF/CC executes this mission through the AKRCC.
U.S. Coast Guard	Recognized SC for all other U.S. aeronautical and maritime SRRs. This includes the State of Hawaii as well as waters over which the U.S. has jurisdiction, such as navigable waters of the U.S.
Other SAR Responsibilities	
National Park Service	NPS provides emergency services on lands and waters administered by NPS, assists visitors within the National Parks or National Monuments, and aids authorities in neighboring jurisdictions.
States and Local Jurisdictions	Outside the above listed SC and Federal civil SAR responsibilities, State and local authorities are responsible for land-based SAR, or may designate a person to be SC (Responsible Authority) within their respective jurisdictions.

The SC can either provide SAR services or make suitable arrangements, with SC oversight for services to be provided by State, Tribal, Territorial, local or other entities.

As previously stated, SCs are responsible for establishing and providing SAR services within a SRR, which are internationally recognized areas of defined dimensions associated with an RCC.

Figure 1-1-1 graphically displays the U.S. SAR Regions.

Each SRR has one 24-hour RCC, and both the SRR and its RCC are recognized by an appropriate international body operating under the umbrella of the United Nations (either the International Maritime Organization (IMO), International Civil Aviation Organization (ICAO), or both). RCCs coordinate aeronautical and maritime SAR operations within their designated SRR as part of the global SAR system, which divides the surface of the Earth into a patchwork of aeronautical and maritime SRRs. Under this system, countries are not responsible for rescuing their own citizens worldwide, but rather the SC, in accordance with treaty provisions, is responsible for providing SAR services within their respective SRR. The SC is required to have provisions in place to search for and rescue persons in distress regardless of their nationality, status, or circumstances.

SAR Mission Coordinator (SMC)

RCCs assign a person, either on a case-by-case basis or on a pre-designated basis, to serve as *SAR Mission Coordinator (SMC)* for each SAR case or incident. SMC responsibility may be transferred as necessary for operations of long duration.

To enable competent and timely response to distress, the SC delegates considerable authority to the SMC to make decisions concerning the SAR operation and ensures that persons who serve as SMC are properly trained and experienced to coordinate and oversee SAR operations.

The SMC is responsible for planning the search and rescue of persons in distress and coordinating transit of SAR resources to the scene.

(Note: Additional information concerning SMC responsibilities is located in the IAMSAR Manual.)

Rescue Coordination Center (RCC)

A Rescue Coordination Center (RCC) is a unit responsible for promoting efficient organization of SAR services and coordinates the conduct of SAR operations within an SRR. The U.S. has established 11 Federal RCCs:

- The Coast Guard operates 9 RCCs;
- U.S. Northern Command operates an RCC for the continental U.S. other than Alaska; and
- U.S. Pacific Command operates an RCC for Alaska.

A *rescue sub-center (RSC)* is a unit subordinate to an RCC established to compliment the RCC according to particular provisions of the responsible authorities. The Coast Guard operates two RSCs (Puerto Rico and Guam).

(Note: Further information on RCCs can be found in Section 1-4.)

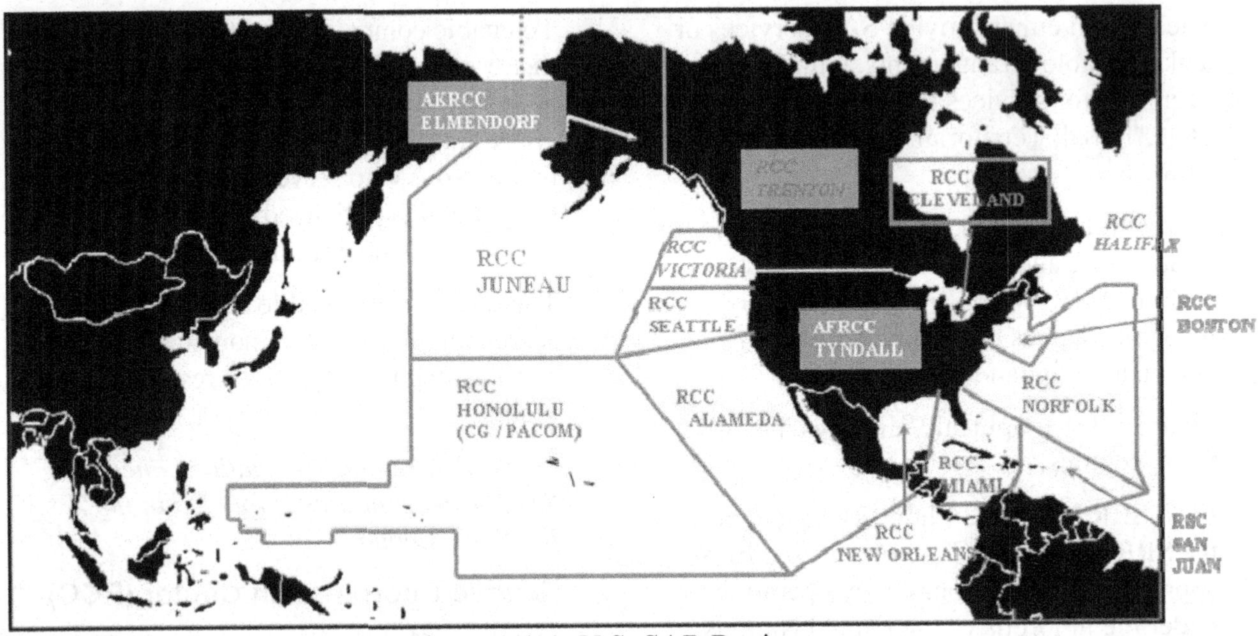

Figure 1-1-1: U.S. SAR Regions

International SAR System

The international SAR system, as identified in the IAMSAR Manual, is implemented through RCCs and is based on responsibilities assigned to a specific SRR. The U.S. SRRs and SAR system is an integral part of the international SAR system. Table 1-1-2 explains the international SAR system general levels, functions, and responsibilities.

Under the international SAR system, SAR resources on scene are typically coordinated by an *On Scene Coordinator (OSC)* supporting the SMC/Incident Commander (IC) by implementing the SMC/IC's SAR Action Plans, coordinating the SAR facilities on scene, ensuring safety, reporting, etc.

For example, when the Coast Guard RCC is notified of a mariner in distress at sea, the Coast Guard will coordinate the SAR operation using the existing international SAR system.

(Note: Additional information on the international SAR system can be found in Section 1-6.)

Table 1-1-2: International SAR System	
Position	**General Functions**
SAR Coordinator (SC)	**Management** Definition: One or more persons or agencies within an Administration with overall responsibility for establishing and providing SAR services and ensuring that planning for those services is properly coordinated (IAMSAR Manual). SC responsibilities include: • Establish, staff, equip and manage the SAR system; • Provide appropriate legal and funding support; • Establish RCCs and RSCs; • Provide or arrange for SAR facilities and SAR resources; • Coordinate SAR training and exercises; and • Promulgate SAR policies and supporting documents.
SAR Mission Coordinator (SMC)	**Mission Planning** Definition: The official temporarily assigned to coordinate the response to an actual or apparent distress situation (IAMSAR Manual). SMC duties include, but are not limited to: • Obtain and evaluate emergency data; • Plot search areas and develop Search Action Plans; • Coordinate activities with the RCC/IC; • Evaluate incoming reports and modify the action plan as appropriate; • Coordinate logistics support for SAR assets; • Coordinate SAR reporting activities; and • Prepare final results report.
On-Scene Coordinator (OSC)	**Operational Oversight** Definition: A person designated to coordinate SAR operations within a specified area. OSC duties include, but are not limited to: • Assume operational coordination of SAR facilities on-scene; • Implement the search action plan; • Keep the SMC advised of any changes to the search action plan; • Develop and implement the rescue plan (if required); and • Make consolidated situation reports to the SMC.

National Incident Management System (NIMS)/Incident Command System (ICS)

While the international SAR system is geographically based and utilizes RCCs, SAR operations conducted using NIMS/ICS, are incident specific and designed for use in any type of emergency response, but especially those involving multiple jurisdictions and authorities

(Note: For additional information concerning NIMS/ICS see Section 1-7).

NIMS/ICS emphasizes:

- A single set of objectives;
- A collective, strategic approach;

- Optimizing information flow and coordination;
- Understanding joint priorities;
- Respecting legal requirements; and
- Maximizing probability of success under a single plan.

When the ICS is implemented, the SMC/IC function will be placed under the umbrella of the NIMS organizational structure. Typically, the SAR Branch Director or SAR Group Supervisor is placed in the Operations Section, where the SAR response system is integrated into the ICS (Figure 1-1-2). The SAR response may also include an OSC and an *Aircraft Coordinator (ACO)* to assist managing critical SAR resources, as required.

In some cases, the person serving as IC may also be designated as the SMC. The terms "Incident Commander" or "Operations Section Chief" are not interchangeable with titles associated with SAR response functions.

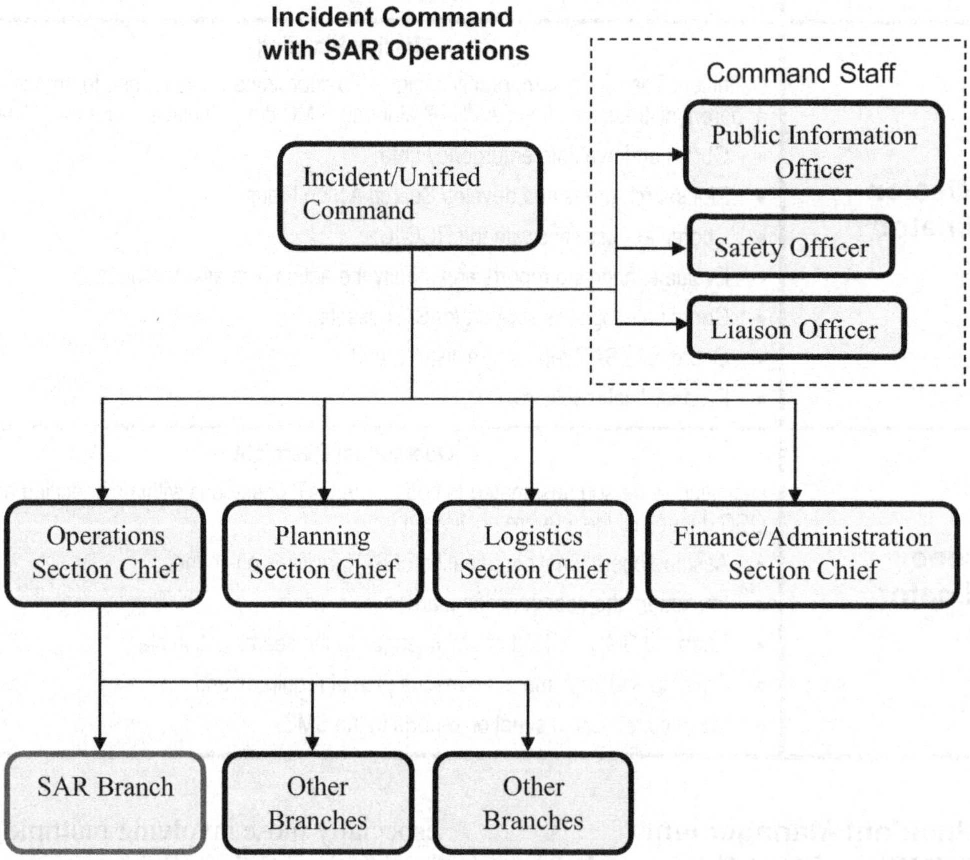

Figure 1-1-2: Incident Command with SAR Operations

General Principles

Because of the number of interagency relationships, types of SAR operations, mix of military and civilian entities, jurisdictions, agreements, emergency response systems involved, traditional practices, and the mix of international and national obligations, land SAR responsibilities can be confusing. In large land SAR operations, SAR authorities may consider implementing an NIMS/ICS Unified Command.

There must be a continuum of responsibilities so there are no gaps in providing SAR services. Generally, aeronautical SAR responsibilities (e.g., the rescue of persons from an aircraft accident/incident) generally flow from the Federal to local levels, whereas land SAR responsibilities generally flow from the local to Federal levels.

There are certain SAR principles that may help in understanding the NSP, National Response Framework (NRF), and related responsibilities among the various levels of government responsible for SAR.

Table 1-1-3 provides general principles for Federal SAR response agencies:

Table 1-1-3: General Principles for Federal SAR Responders
1. Persons in distress are assisted without regard to their nationality, status or circumstances;
2. The governing and guiding SAR reference documents generally apply in descending order of precedence from international to local, but take into account National and State sovereignty;
3. Jurisdictional and lifesaving concerns must be balanced (political, economic, jurisdictional, or other such factors should normally remain secondary when conducting lifesaving operations);
4. SAR coordination and other services can be either provided or arranged;
5. SAR responsibilities can be delegated but not relinquished;
6. SAR plans and services in the U.S. emphasize cooperation rather than legislation;
7. SAR responsibilities under the NSP are generally based on geography, while services under the NRF are generally incident specific;
8. Rather than establishing SRRs, the NSP recognizes already established international aeronautical SRRs for land SAR;
9. SRRs are established to help ensure adequate provision of SAR services and are not intended to obstruct prompt assistance to persons in distress;
10. SCs are responsible for civil SAR identified in the NSP within their respective SRRs or other areas of responsibility (this includes any Federal role and also appropriate arrangements and coordination with Federal, State, Tribal, Territorial, or local authorities);
11. In accordance with the NSP, SCs should delegate authority to their RCC(s) to plan and coordinate SAR operations in recognition of the extensive expertise required for SAR and its extremely time-critical nature;
12. SAR relationships and responsibilities can be re-aligned or clarified by various types of agreements, as long as these agreements are consistent with higher level policies, guidance, and applicable law, and the parties are the proper authorities that would be responsible for the issues identified;
13. Provisions of the NSP are intended to help fulfill both legal and humanitarian obligations;
14. Arrangements between Federal, military, and civil agencies should provide for the fullest practicable cooperation, consistent with statutory authorities and responsibilities assigned civil SAR functions;
15. Aeronautical and maritime SAR in the U.S. internationally recognized SRRs must be organized and carried out in accordance with provisions that apply to the international SAR system;
16. The provisions of NIMS/ICS are also applied as warranted by the nature of the operations and normal practices of the authorities involved in an emergency response; and
17. Absent alternative arrangements, authorities fulfilling SAR responsibilities under the NSP generally do so at their own expense, and those fulfilling SAR obligations under Emergency Support Function (ESF) #9 can generally be reimbursed.

In addition, Table 1-1-4 provides general principles for States, Tribes, Territories, and local jurisdictions in the coordination of SAR services:

Table 1-1-4: General Principles for States, Tribes, Territories, and Local Governments Concerning SAR Responsibilities
1. States, Tribes, Territories, and local jurisdictions are not required to comply with the NSP.
2. When States request Federal SAR assistance, they should coordinate that assistance through prearranged agreements;
3. States, Tribes, Territories, and local jurisdictions are normally not involved in actions necessary to comply with international treaties, guidance, and agreements related to SAR;
4. States, Tribes, Territories, and local jurisdictions will generally retain SAR responsibilities within their jurisdictions for incidents primarily local or intrastate in character. In such cases, appropriate agreements are generally made between Federal SCs and relevant State authorities to coordinate the use of requested Federal assistance;
5. States, Tribes, Territories, and local Jurisdictions are generally responsible for land-based SAR. States normally accept these responsibilities through legislation, plans, agreements and other means, and are encouraged to designate a person to serve as State SC to oversee State SAR services and coordinate with Federal SCs; and
6. States should develop cooperative arrangements with other authorities (Federal, military, State, Tribal, Territorial, local and volunteer SAR responders) to ensure that SAR services can be effectively provided within their jurisdiction.

Types of Civil SAR

The NSP identifies the following types of civil SAR:

- Maritime (involving rescue from a water environment);
- Aeronautical (including civil SAR assistance in the vicinity of airports);
- Land (including civil SAR operations associated with environments such as remote areas, swift water, caves, mountains, etc.);
- Urban SAR (US&R);
- Provision of initial assistance at or near the scene of a distress situation (e.g., initial medical assistance or advice, medical evacuations, provision of needed food or clothing to survivors, etc.);
- Delivery of survivors to a place of safety or where further assistance can be provided, or further transportation arranged if necessary;
- Saving of property when it can be done in conjunction with or for the saving of lives;
- Mass rescue operations (MROs); and
- Catastrophic incident SAR (CISAR).

Terminology and Acronyms

The best reference for SAR terms, definitions, and acronyms is the NSS; more limited entries are provided at the beginning of this Addendum.

It is intended that all terms used in either the NSS or any of its Addendums be included in the NSS's glossary and list of acronyms so that the NSS can serve as the primary source for this information.

The IAMSAR Manual provides an extensive glossary and list of acronyms as well, but most of that information is also included in the NSS. U.S. Federal SCs follow international SAR provisions and use terminology accepted worldwide, as practicable.

In addition to these generally accepted terms and definitions, relevant directives of the Federal Agencies with SAR responsibilities also include Agency-specific relevant terms and acronyms.

Saving Property

SAR authorities are not required to "rescue" property; such operations are always discretionary and secondary to lifesaving and any other higher priority operation.

This page intentionally left blank.

Section 1-2: Key References

SAR doctrine, policy and procedures for Federal SAR agencies are provided in primary publications that broadly apply to civil SAR, and which may be inter-related or supplemented.

Since these publications are discussed thoroughly within the NSP, NSS, Addenda to the NSS and other places, this Section provides only general information.

International Treaties, Conventions, and Agreements

There are numerous international treaties and agreements that apply to civil SAR within the U.S. and with neighboring countries. The primary international convention of interest to land SAR authorities is the Convention on International Civil Aviation (Annex 12 - Search and Rescue) applies to persons in distress in both domestic and international flights.

International Aeronautical and Maritime Search and Rescue (IAMSAR) Manual

The IAMSAR Manual provides guidance on aeronautical and maritime SAR applicable both internationally and nationally. For land SAR, the IAMSAR manual applies from an aeronautical SAR perspective.

The IAMSAR Manual consists of three volumes:

- *Volume 1 (Organization and Management)* discusses the:
    - Global SAR system concept;
    - Establishment and improvement of national and regional SAR systems;
    - Cooperation with neighboring States to provide effective and economical SAR services; and details SC responsibilities.
- *Volume 2 (Mission Coordination)* provides guidance to personnel who plan and coordinate SAR operations and focuses on information for RCC personnel and SMCs.
- *Volume 3 (Mobile Facilities)* is intended to be carried aboard rescue units, aircraft, and vessels to help with performance of a search, rescue, or OSC function.

The IAMSAR manual is available in multiple languages and available for purchase from various sources, including the IMO and ICAO websites.

National SAR Plan (NSP)

The NSP is a Federal cabinet-level inter-agency document that describes how the U.S. will meet its international legal and humanitarian obligations to provide SAR services. It establishes over-arching Federal SAR policy, assigns SAR responsibilities to Federal Agencies, and adopts the IAMSAR Manual and the NSS for use by U.S. SAR agencies. The NSP, NSS and other Federal SAR documents can be found at the NSARC website: www.uscg.mil/nsarc. A copy of the NSP is provided in appendix I.

National SAR Supplement (NSS)

The NSS, which is complemented by this and other Addenda, provides implementing policy guidance for the NSP. It also describes the national SAR system, expands on topics covered by the IAMSAR Manual, and provides guidance that applies uniquely for the U.S.

National Response Framework (NRF)

The NRF is developed under the authority of the Robert T. Stafford Disaster Relief and Emergency Assistance Act ("Stafford" Act) and at the direction of the President. It has 15 ESF Annexes; ESF #9 concerns SAR.

The NRF is the Federal plan-of-action for responding to disasters that meet the following criteria:

- The State and local response capabilities are overwhelmed;
- The State government requests Federal assistance; and
- The President formally declares that a disaster has occurred, activating the disaster assistance authority outlined in the Stafford Act (42 USC. § 5121 et seq.).

Once these criteria have been met, the Federal Government may implement any or all of the 15 ESFs that are described in the NRF. Guidance for implementing ESF #9 is provided in the CISAR Addendum (available for download at the NSARC website). The NRF and its ESF Annexes are available at: www.fema.gov/emergency/nrf/ ESF #9 can also be found in appendix J.

Summary

International SAR-related treaties, conventions, and agreements to which the U.S. is Party, and their implementing guidance (e.g., the IAMSAR Manual), apply to U.S. civil SAR operations.

The NSP and its implementing guidance (e.g., the NSS and its Addenda) not only expands on international obligations and guidance, but is broader in scope, establishing the U.S. national SAR system as an integral part of the global SAR system.

Section 1-3: Federal SAR Authorities

Federal Inland SAR Coordinator (SC)

Aeronautical SAR

Key Federal Land SAR Agencies and Organizations

- *National SAR Committee (NSARC)*
- *Department of Defense (DoD)*
- *U.S. Coast Guard (USCG)*
- *Federal Emergency Management Agency (FEMA)*
- *National Park Service (NPS)*

Other Relevant Federal Agencies

- *U.S. Forest Service*
- *National Oceanic and Atmospheric Administration (NOAA)*
- *Customs and Border Protection (CBP)*
- *Immigration and Customs Enforcement (ICE)*
- *Bureau of Indian Affairs*
- *Department of State (DOS)*
- *Federal Aviation Administration (FAA)*
- *Federal Communications Commission (FCC)*
- *National Aeronautics and Space Administration (NASA)*
- *National Transportation Safety Board (NTSB)*

Federal Land SAR Coordinator (SC)

CDRUSNORTHCOM is the recognized Federal SC for the aeronautical SRR corresponding to the continental U.S. other than Alaska.

This does not mean that CDRUSNORTHCOM, as the Federal SC for the continental U.S., is only responsible for aeronautical SAR (assistance to persons in distress in aircraft). As a Federal SC, CDRUSNORTHCOM is responsible for assisting State SCs/Responsible Authorities and providing SAR coordination among multiple agencies and authorities within the continental U.S. (Langley SRR), which is internationally recognized as an aeronautical SRR.

Why does the NSP identify the continental U.S. SRR as an *aeronautical* SRR?

Internationally, there are two primary conventions that govern international SAR:

- *IMO's International Convention on Maritime Search and Rescue (1979)*; and
- *ICAO's Convention on International Civil Aviation (1949) Annex 12 (Search and Rescue).*[1]

There is no international regime that governs land-based SAR within a nation's sovereign territory.

As a result, CDRUSNORTHCOM's AOR is internationally recognized through the Chicago Convention, which governs the implementation of aeronautical SAR.

Aeronautical SAR

This Addendum focuses on land SAR coordination, procedures, search planning, and rescue operations. For aeronautical SAR that may occur over land, the governing convention is the Chicago Convention (Annex 12 - Search and Rescue). Other documents that concern aeronautical SAR include the:

- ICAO North America Regional Air Navigation Plan (RANP);
- IAMSAR Manual;
- NSP;
- NSS; and other
- Addenda to the NSS.

The ICAO-recognized U.S. aeronautical SRRs are described in the NSS. The respective RCCs for each SRR are responsible not only for aeronautical SAR, but for any appropriate Federal involvement in any type of civil SAR covered by the NSP.

The Chicago Convention (Annex 12) applies to SAR services for persons in distress in any type of aircraft, not just aircraft that are subject to the other provisions of the Convention (i.e., aircraft on international flights). Therefore, coordination and provision of any aeronautical SAR services, regardless of the type of aircraft involved, is the responsibility of the U.S. Government as a "Contracting State" to the Convention. The U.S. fulfills these international obligations by assigning SC responsibilities in the NSP. Treaty provisions may be further limited or expanded by international agreements to which the U.S. is Party.

While Federal SCs can arrange via agreements for States to coordinate aeronautical SAR, and they may be directly involved in responding to an aeronautical incident on an infrequent basis, the overall international responsibility for ensuring that aeronautical SAR services are provided as required in international standards, lies with the Federal SCs.

Regardless of the State, Tribal, Territorial, or local SAR authorities involved, a Federal RCC should normally coordinate aeronautical SAR operations that involve:

- Overdue general aviation (GA) aircraft on an interstate flight;
- DoD aircraft in distress; and
- Commercial aircraft on missions of national concern (e.g., National Special Security Events, Special Events for Homeland Security, etc. During these events, the Domestic Event Network (DEN) is the controlling authority.).

[1] Also known as the "Chicago" Convention.

Key Federal Land SAR Agencies and Organizations

National SAR Committee (NSARC)

NSARC oversees matters at the Federal level related to the NSP. NSARC Member Agencies include the following:

- Department of Commerce (DOC);
- Department of Defense (DoD);
- Department of Homeland Security (DHS);
- Department of Interior (DOI);
- Department of Transportation (DOT);
- Federal Communications Commission (FCC); and
- National Aeronautics and Space Administration (NASA).

General responsibilities of NSARC Member Agencies are described in the NSP and NSARC agreement.

The following sections provide a brief overview of NSARC Member Agencies, as well as other Federal Agencies that assist in the conduct of land SAR.

Further information concerning the National SAR Committee can be found at: www.uscg.mil/nsarc.

Department of Defense (DoD)

Based on the NSP and NRF, the Secretary of Defense directs all Services and other components of the U.S. Department of Defense to support civil SAR services, including CISAR. As established by the NSP, DoD's roles in civil SAR are carried out on a not-to-interfere basis with DoD's primary military duties.

DoD has capabilities that are very suitable for land SAR operations, from a single helicopter rescue to large-scale MROs and CISAR.

DoD operates two internationally-recognized RCCs for civil SAR:

- RCC Tyndall (AFRCC), Tyndall Air Force Base (AFB), Panama City, Florida, for SAR operations in the continental U.S. other than Alaska under CDRUSNORTHCOM acting as the SC for the Langley SRR; and
- RCC Elmendorf (AKRCC), Elmendorf AFB, Anchorage, Alaska, for SAR operations in Alaska under CDRUSPACOM acting as the SC for the Elmendorf SRR.

Two key DoD references concerning civil SAR include:

- DoDD 3025.18, Defense Support of Civil Authorities (DSCA); and
- DoDI 3003.01, DoD Support to Civil SAR.

DSCA refers to DoD support provided by Federal military forces, DoD civilians and contract personnel, and DoD agencies and components, in response to requests in anticipation of, and for assistance from civil SAR authorities for emergencies, law enforcement support and other domestic activities, or from qualifying entities for special events. Requests for DSCA come through established procedures from recognized civil authorities.

CDRUSNORTHCOM has responsibility for homeland defense (HD) and providing DSCA within its assigned AOR, including Alaska. CDRUSPACOM has responsibility for HD, and DSCA for Hawaii, U.S. Territories, and other areas that may be assigned within its AOR.

More DSCA information is available in the CISAR Addendum to the NSS.

Both DoD and DOI/NPS are NRF ESF #9 Primary Agencies for Land SAR when a State requests Federal SAR resources for disaster operations.

U.S. Coast Guard (USCG)

The USCG, operating under the Department of Homeland Security (DHS), is the Federal SC of the U.S. aeronautical and maritime SRRs in the oceanic environment, the Great Lakes, and all U.S. navigable waters. USCG SAR policy is located in the Coast Guard Addendum to the NSS.

USCG conducts land and CISAR operations as required.

Under the NRF's ESF #9, USCG is the Primary Agency for Maritime/Coastal/Waterborne SAR which includes operations for natural and man-made disasters that primarily require USCG air, ship, boat, and response team operations, and supports unified SAR operations conducted in accordance with the NSP.

USCG sponsors and chairs the NSARC which oversees the national SAR system and U.S. involvement in international SAR.

Additional information on Coast Guard SAR is available at the Office of Search and Rescue, Coast Guard Headquarters website: http://www.uscg.mil/hq/g-o/g-opr/G-OPR.htm .

Federal Emergency Management Agency (FEMA)

FEMA, operating under the Department of Homeland Security (DHS), is the Federal coordinating agency (ESF Coordinator) for the NRF's ESF #9. FEMA coordinates Federal resources during CISAR and may be involved in SAR operations not covered by the NRF.

FEMA is an ESF #9 Primary Agency with responsibility under the NRF for urban SAR (US&R).

US&R is structural collapse SAR, and includes operations for natural and man-made disasters and catastrophic incidents as well as other structural collapse events that require SAR operations.

FEMA coordinates Federal US&R planning activities, and each of the ten FEMA regions produce supplemental response plans, including sections on US&R, based upon known resources, capabilities, and State,

Tribal, Territorial, and local authorities in their areas of responsibility.

FEMA develops national US&R policy, provides planning guidance and coordination assistance, standardizes task force procedures, evaluates task force operational readiness, funds special equipment and training within available appropriations, and reimburses, as appropriate, task force costs incurred as a result of ESF #9 deployment.

FEMA activates the National Urban Search and Rescue Response System for incidents requiring a coordinated Federal US&R response. Upon activation under the NRF, US&R task forces are considered Federal assets under the Robert T. Stafford Disaster Relief and Emergency Assistance Act and other applicable authorities. The System integrates US&R task forces, Incident Support Teams (ISTs), and technical specialists in support of unified SAR operations conducted in accordance with the NSP. The task forces support State, Tribal, Territorial, and local response efforts to locate survivors and manage recovery operations.

FEMA reimburses for authorized US&R deployments to Stafford Act declaration sites, but does not have the authority or funding to reimburse activities absent a Stafford Act declaration. Non-Stafford Act US&R deployments are reimbursed by the Federal authority requesting US&R assistance in accordance with provisions contained in the Financial Management Support Annex to the NRF.

Further information on FEMA US&R is available at: http://www.fema.gov/emergency/usr.

National Park Service (NPS)

NPS, operating under the Department of Interior (DOI), provides SAR services on lands and waters administered by NPS and aids authorities in neighboring jurisdictions. Civil SAR operations, including emergency medical aid, are conducted in a wide variety of environments such as remote, rural and roadless areas, lakes, rivers and oceans, and deserts, mountains and caves, and often require extended response times and use of specialized equipment. NPS works closely with Federal, State, Tribal, Territorial, and local SAR organizations.

National Parks have established local SAR plans that provide park managers with direction and guidance in establishing and managing SAR operations.

NPS provides visitor protection services, including varying levels of SAR, and provides aid and assistance to visitors, which may include search, rescue, and medical assistance.

Qualified SAR services in local communities may be utilized if they can provide a timely response to SAR incidents within an NPS area. When such services are not available, the NPS will make a reasonable effort to provide a level of SAR service commensurate with park needs.

NPS, Branch of Emergency Services, provides oversight, direction, and coordinates the Service-wide SAR Program.

NPS considers search to involve finding lost persons in a wide variety of environments from buildings to high mountains and transporting them to safety, and considers rescue to involve accessing, stabilizing, extricating and transporting stranded or injured persons using available resources ranging from hand-carried litters to hoist-capable helicopters.

NPS cooperates with and supports State, Tribal, Territorial, and local governments where its units are located. States may not regulate the NPS without specific congressional consent; however, park areas may adopt all or part of the State SAR policies and guidelines as long as they don't conflict with NPS Director's Orders.

NPS, along with DoD, is a Primary Agency under the NRF's ESF #9 responsible for land SAR. Other DOI assets will be coordinated and utilized to enhance the NPS's capabilities.

Further information on the NPS is available at: http://www.nps.gov/index.htm.

Other Relevant Federal Agencies

SAR sometimes requires the services of other Federal agencies. While each agency's involvement in SAR varies with their primary mission and availability of resources, all are integral to the National SAR System.

U.S. Forest Service (USFS)

USFS, operating under the U.S. Department of Agriculture (USDA), manages the national forests and grasslands. USFS has law enforcement personnel and other staff that help ensure the safety of persons using national forests and who respond to emergencies in those locations, and has resources applicable to specific areas throughout the country.

USFS has agreements with local authorities in many areas to facilitate mutual assistance.

Further information on the USFS is available at: http://www.fs.fed.us/.

National Oceanic and Atmospheric Administration (NOAA)

NOAA, operating under the Department of Commerce (DOC), is the lead U.S. agency for U.S. Search and Rescue Satellite Aided Tracking (SARSAT) Program, which is a component of the International Cospas-Sarsat Programme. NOAA operates the U.S. Mission Control Center (USMCC) as part of the U.S. SARSAT program ground system, and is responsible for maintaining the US 406 MHz Beacon Registration Database. The Cospas-Sarsat system automatically forwards distress beacon alerts and registration data to RCCs and other SAR Points of Contact (SPOCs).

Information on the Beacon Registration Database is available at: http://www.beaconregistration.noaa.gov.

Information on the U.S. SARSAT program is available at: http://www.sarsat.noaa.gov/usmcc.html.

Information on the International Cospas-Sarsat Programme is available at:

http://www.cospas-sarsat.org/index.php.

Customs and Border Protection (CBP)

CBP operates under DHS.

Border Patrol Search, Trauma, and Rescue (BORSTAR) Teams are highly specialized units capable of responding to emergency search and rescue situations anywhere in the United States. These teams are made up of Border Patrol agents who volunteer to go beyond their regular duties and undergo a highly specialized regimen consisting of training in physical fitness and other disciplines, including medical skills, technical rescue, navigation, communication, swiftwater rescue, and air operations.

BORSTAR's primary mission is to respond to incidents involving distressed agents and migrants along the border. These rescued individuals are predominantly undocumented aliens but also include agents and border residents. SAR operations are conducted throughout the year in varying climates and topographies. BORSTAR agents have aided in rescue operations lasting up to 5 days. These rescues vary in difficulty from locating distressed persons to complex rescue operations in remote locations.

Immigration and Customs Enforcement (ICE)

U.S. Immigration and Customs Enforcement

ICE is an investigative arm of DHS responsible for eliminating vulnerabilities in the nation's border and its economic, transportation and infrastructure security.

ICE is composed of four law enforcement divisions and several support divisions that provide investigation, interdiction and security services to the public and law enforcement partners in the Federal and local sectors. Several ICE units have helicopters and fixed wing aircraft that may be used in searches.

ICE can help to determine if aircraft cross or intend to cross the US border.

Further information on ICE is available at: http://www.ice.gov/.

Bureau of Indian Affairs (BIA)

The BIA, which operates under DOI, manages land held in trust by the U.S. for American Indians, Indian tribes, and Alaska Natives. There are over 560 Federally recognized tribal governments in the U.S. These tribal lands are considered domestic dependent nations by the Federal Government.

Local tribal governments are responsible authority for SAR on tribal lands. Responsible tribal authorities can request assistance from Federal, State, other Tribal, and local SAR authorities.

To request ESF #9 SAR assistance under the NRF, Tribes shall be included in a State's request for Federal disaster assistance. After a disaster declaration has been approved, the Tribe(s) may work directly with the Federal Government to obtain SAR resources and assistance.

Further information on the BIA is available at: http://www.BIA.gov/.

Department of State (DOS)

DOS designates Federal Agencies responsible for civil SAR to represent the U.S. in the following international forums:

- The USCG leads and coordinates U.S. participation in SAR-related initiatives at the IMO;
- The FAA, with support of NSARC Member Agencies, leads and coordinates U.S. participation in SAR- related initiatives at ICAO;
- NOAA leads and coordinates U.S. participation in the International Cospas-Sarsat Programme and associated international programs;
- The USAF and USCG each provide a SAR expert to serve as a member of the ICAO-IMO Joint SAR Working Group; and
- United States Agency for International Development (USAID)/Office of Foreign Disaster Assistance (OFDA) leads and coordinates national participation in international urban SAR activities, particularly with the United Nations.

Further, as discussed in the NSP, any SAR-related international treaty or agreement or any type, or its implementation, is subject to DOS oversight.

Further information on DOS is available at: http://www.state.gov/.

Federal Aviation Administration (FAA)

The FAA, operating under the Department of Transportation (DOT), through its Air Route Traffic Control Centers (ARTCC) and Flight Service Stations (FSS), monitors and flight-follows aircraft and may be the first agency to alert an RCC of an aviation emergency or overdue aircraft.

Once alerted, the RCC and FAA work together to locate the aircraft, reviewing radio communications (if available) and radar data to ascertain as closely as possible a good last known position (LKP). Concurrently, other FAA facilities conduct "ramp" checks at airports where an aircraft may have landed.

The FAA can recall recorded radar data and identify and trace aircraft that are at a sufficient altitude to be tracked by radar. RCCs can seek this radar data from the FAA, which can greatly assist in aircraft searches, as well as providing route and last radar position data.

Most aircraft are required by the FAA to carry emergency locator transmitters (ELTs) that are designed to automatically activate in the event of a accident.

Further information about the FAA and its SAR capabilities is available at: www.faa.gov/air_traffic/publications and FAA Order JO 7110.10 series (Flight).

Federal Communications Commission (FCC)

The FCC, an independent Federal Agency established by the Communications Act of 1934, regulates interstate and international communications by radio, television, wire, satellite, and cable throughout the U.S. FCC regulations cover communications equipment and frequencies used for distress alerting and SAR operations, including some vessel and aircraft radio licensing information that may be helpful to SAR personnel (however, distress beacon registration information is in a database maintained by NOAA). The FCC has statutory administrative authority for communications violations, including radio-related abuse violations such as repeat false alerts or hoaxes. The FCC can also provide radio direction-finding support.

RCCs work closely with the FCC to report unauthorized use of emergency frequencies, which may be subject to administrative actions (including forfeitures) by the FCC under Title 47, United States Code. Violators of the FCC rules and regulations may also be subject to imprisonment of up to one year and/or seizure of the radio equipment. In addition, Title 14 U.S.C. 88(c) (under United States Coast Guard) makes it a Federal felony, punishable by significant imprisonment and/or a monetary fine, for anyone who knowingly and willfully communicates a false distress signal, and the FCC has been a supporting agency in past Coast Guard actions.

Further information on the FCC is available at: http://www.fcc.gov/.

National Aeronautics and Space Administration (NASA)

NASA is an independent Federal Agency responsible under the NSP with supporting SAR through research and development or application of technology to search, rescue, survival, and recovery systems and equipment, such as location tracking systems, transmitters, receivers, and antennas capable of locating aircraft, ships, spacecraft, or individuals in potential or actual distress.

Further information on NASA is available at: http://searchandrescue.gsfc.nasa.gov/.

National Transportation Safety Board (NTSB)

The NTSB is an independent Federal Agency charged by Congress with investigating and determining the probable causes of every civil aviation accident in the U.S. and significant accidents in the other modes of transportation (e.g., railroad, highway, marine and pipeline) and issuing safety recommendations aimed at preventing future accidents.

NTSB maintains a Go Team that begins investigation of a major accident at the accident scene as quickly as possible, assembling a broad spectrum of technical expertise to solve complex transportation safety problems.

NTSB maintains an accident database that may help identify previous wreckage locations.

Further information on the NTSB is available at: http://www.ntsb.gov/

Section 1-4: Rescue Coordination Centers (RCCs)

Overview

An RCC is an operational facility responsible for promoting efficient organization of SAR services and for coordinating the conduct of SAR operations within an SRR. Each RCC coordinates SAR facilities in the conduct of SAR operations in their respective SRR.

601st Air Operations Center, Tyndall Air Force Base, Florida. The Air Force Rescue Coordination Center is embedded within "America's AOC".

An RCC that coordinates both aeronautical and maritime SAR is internationally recognized as a Joint RCC (JRCC). In the continental U.S., the AFRCC and AKRCC are both internationally recognized as Aeronautical RCCs (ARCCs), even though they both coordinate the conduct of land SAR operations.

(Note: Coast Guard RCCs are JRCCs because they coordinate both aeronautical and maritime SAR operations in the U.S. oceanic SRRs.)

RCC personnel should be familiar with the capabilities of available SAR facilities in their respective SRR. Collectively, these SAR facilities are the means by which the RCC conducts its SAR operations. Some of these SAR facilities will be immediately suitable for use; others may have to be enhanced by changing organizational relationships or supplying extra equipment and training. If the SAR facilities available in certain parts of an SRR cannot provide adequate assistance, arrangements should be made to provide additional facilities.

RCC Requirements

Basic requirements for RCCs include:

- 24-hour availability;
- Trained personnel;
- Charts and maps which apply to the RCC's SRR;
- Means of plotting;
- Ability to receive distress alerts from equipment normally used for alerting within the SRR (e.g., Mission Control Centers (MCCs), air traffic services (ATS), cell phones, and other alerting posts, etc.);
- Immediate communications with associated FAA facilities and other RCCs;
- Rapid and reliable communications with parent agencies of SAR facilities, meteorological offices, and SAR facilities conducting SAR operations;
- Phone, fax, computers, software, email, databases, and any other such tools that are needed for efficient and effective RCC operations; and
- Plans of operation.

Additional information that should be available to RCC personnel includes:

- Applicable international, national, State and other useful SAR publications;
- Air navigation regulations;
- Communications publications;
- Aeronautical information publications (AIPs);
- Indexes of names, addresses, telephone and facsimile numbers; and
- Relevant checklists and forms.

RCC personnel must be trained and capable of planning and coordinating the volume and types of SAR operations that may occur within the SRR. If the RCC staff has duties besides SAR, the additional functions should be considered when determining staffing requirements. An RCC must be in a constant state of operational readiness. Where the RCC does not maintain sufficient continuous staffing on duty, provision must be made for stand-by RCC staff to be rapidly mobilized.

An SMC should be designated (may be pre-designated) for each specific SAR operation, and adequate numbers of personnel qualified to perform the SMC function must be readily available on a 24-hour basis. This is a temporary function that may be performed by a designated SAR duty officer, assisted by as many staff as may be required.

A SAR operation may continue over a prolonged period of time. The SMC is in charge of a SAR operation until a rescue has been concluded or until it has become apparent that further efforts would be of no avail.

RCC personnel should be trained, qualified, and certified to coordinate SAR operations and effectively stand the RCC watch. Qualification and certification processes ensure that RCC personnel have the prerequisite experience, maturity, and judgment to coordinate SAR operations. During the qualification process, RCC personnel must, by demonstration of abilities, show mental and physical competence to perform as part of a team. RCC personnel certification is official recognition by the SAR organization that it trusts the person in the conduct and coordination of SAR operations.

Each RCC must prepare comprehensive plans of operation for the conduct of SAR in its SRR. These plans must cover potential SAR scenarios for the whole SRR and be based on agreements between authorities responsible for the RCC and the providers of SAR facilities or other support for SAR operations.

RCC Information Sources

An RCC should possess investigative capabilities to assist in all phases of a SAR event.

Telephone investigations. Telephonic communications search may resolve SAR events before alerting and dispatching a SAR facility. Such investigations should attempt to:

- Establish and maintain communications with the Reporting Source(s);
- Locate missing persons, aircraft and/or vessels;
- Identify where aircraft and/or vessels have or have not departed from or transited;
- Determine intentions of missing persons; and
- Search for and validate leads and clues in an attempt to establish disposition of objective and narrow the search area.

Internet investigations. The internet is a tool for investigating incidents. When information such as a tail or hull number, telephone numbers, and the names of missing persons are available, RCC personnel may find information on the internet to help coordinate a suitable response.

SARSAT alert correlation. Distress beacon alert correlation can help RCC personnel transition from the Alert Phase to the Distress Phase during a potential SAR incident.

Cell phone forensics. Agreements and standard operating procedures with cellular service providers can be used to obtain needed information. Cellular service providers and other sources will typically assist SAR authorities in locating the origin of cell calls.

(Note: 18 USC § 2702 (b)(8) states that, "A provider... may divulge the contents of a communication to a Federal, State, or local governmental entity, if the provider, in good faith, believes that an emergency involving danger of death or serious physical injury to any person requires disclosure without delay of communications relating to the emergency.")

Radar forensics. During aircraft missions, radar forensics can facilitate locating the search object, usually by providing a LKP. The RCC should have agreements with sources of radar information. Radar requests should include the relevant date and time of the incident, transponder code, LKP, altitude, flight route, and aircraft type.

Air Defense Sectors (ADS). The Western Air Defense Sector (WADS) and Eastern Air Defense Sector (EADS) have access to radar sites not normally used by the FAA; data is retained for 365 days.

National Alert Warning System (NAWAS). NAWAS is a special purpose telephone system that is a component of the Federal Civil Defense Warning System. NAWAS selectively disseminates warning and emergency information within FEMA regions/States to Federal, State, and other law enforcement agencies using public service systems. Often police, highway patrol, and county sheriffs are linked to State networks and can assist in collecting and broadcasting information. RCCs may issue, through FEMA, a NAWAS request for assistance from law enforcement agencies, subject to State NAWAS restrictions.

Aircraft Wreckage Locator Database (AWLD). The AFRCC maintains an AWLD with all reported aircraft wreckage sites within the continental U.S. The AWLD is an important source that can identify aircraft wreckage already in a search area. The file contains geographic location, wreckage description (including aircraft type and color) and crash date.

The AFRCC works closely with the NTSB to determine the disposition of wreckage after a SAR mission. If the NTSB, aircraft owners, and insurance companies decide to leave the wreckage in place, then data is added to the database. The AWLD is accessible at:

https://1afnorth.region1.ang.af.mil/AFRCC/default.aspx.

Inland RCCs

The U.S. has three RCCs that have land SAR responsibilities. These three RCCs are:

- RCC Tyndall (AFRCC),
- RCC Elmendorf (AKRCC), and
- RCC Honolulu.

RCC Tyndall (AFRCC)

RCC Tyndall (AFRCC) is co-located with the 601st Air and Space Operations Center (601 AOC) at Tyndall Air Force Base, Panama City, Florida. The AFRCC coordinates SAR services for aeronautical and other types of land-based SAR within the aeronautical SRR corresponding to the continental U.S. other than Alaska. The AFRCC seeks to encourage a cooperative SAR network with other Federal, State, Tribal, Territorial and local SAR authorities to help coordinate assistance for persons in distress.

The AFRCC is chartered with the responsibility of "continuously building a coordinated search and rescue network ensuring timely, effective lifesaving operations whenever and wherever needed through cooperation, coordination, and education."

AFRCC's mission is derived from ICAO's Chicago Convention and the IAMSAR Manual. These international requirements are satisfied via the NSP and NSS, providing the U.S. with a national plan for coordinating SAR services to meet U.S. domestic needs and international commitments. Further implementing DoD guidance is provided in Department of Defense Directive (DODD) 3003.01, DoD Support to Civil Search and Rescue.

The AFRCC has no SAR command or control authority, but supports domestic civil authorities by coordinating DoD civil SAR services to the fullest extent practicable on a non-interference basis with primary military duties, and according to applicable national directives, plans, guidelines, and agreements. No DoD resources are assigned or tasked for aeronautical civil SAR. The AFRCC has no tasking authority, SAR facilities, or other assigned SAR resources.

AFRCC is the SMC for incidents and missions involving:

- Overdue GA interstate flights;
- DoD aircraft;
- Commercial aircraft; and
- Missions of national concern (e.g. National Special Security Events - NSSEs).

As SMC, the AFRCC operates with the full operational authority of the Federal SC for the continental U.S. (Commander, U.S. Northern Command).

AFRCC receives all:

- Aircraft ELT and personal locator beacon (PLBs) distress beacon alerts *registered* in the U.S.;
- Any distress beacons registered in other countries which are geographically located and activated in the Langley SRR (continental U.S.); and
- All distress beacon alerts that are U.S. registered, but are geographically located outside the U.S.

If located in the continental U.S., the AFRCC works closely with other SAR authorities to:

- Respond to aeronautical and ground distress beacon reports;
- Investigate the reports in coordination with Federal, State, Tribal, Territorial, and local SAR authorities; and
- Determine the required type and scope of response.

Through SAR agreements with States, the AFRCC may be a coordinating authority for specific land SAR events other than aeronautical. These events may include patient transport, search, rescue, and SAR unit transport.

Each State in the contiguous U.S. has SAR agreements on file with AFRCC to delineate a responsible SC/SAR coordinating agency and coordinating requirements for all SAR missions.

The AFRCC coordinates Federal SAR capabilities at the request of other designated RCCs, as well as other Federal and State SCs in support of ongoing civil SAR operations.

AFRCC can be contacted at:

- Toll Free: 1-800-851-3051;
- Local (both emergency and non-emergency): 850-283-5955.
- Email: afrcc.console@tyndall.af.mil

RCC Elmendorf – Alaska (AKRCC)

The Elmendorf RCC (also called the Alaska Rescue Coordination Center; AKRCC) is associated with the 611th Air and Space Operations Center (611 AOC) at Elmendorf AFB, Anchorage, Alaska, and is administratively referred to as the 11th Air Force RCC. The AKRCC coordinates SAR services for aeronautical and other types of land-based SAR within the aeronautical SRR corresponding to Alaska. The AKRCC also coordinates SAR services for military SAR as well as DoD support to any other civil SAR providers within the Alaskan AOR. AKRCC will assume SMC for incidents involving:

- DoD assets;
- Any aviation incident; or
- When asked by another civil SAR provider (USCG, National Parks Service, or Alaska State Troopers) when their resources or capabilities have been exceeded.

The AKRCC seeks to encourage a cooperative SAR network with the other Federal, State, Tribal, Territorial, and local SAR authorities to help coordinate assistance for persons in distress.

Federal SC responsibility for Alaska has been delegated from CDRUSPACOM to CDRPACAF to Commander, of 11th Air Force (11 AF/CC), which coordinates SAR operations through the AKRCC. AKRCC is a unit of the Alaska Air National Guard

(ANG), manned entirely by AK guardsmen with an AK ANG administrative chain of command and an active duty, 11th Air Force, operational chain of command.

Like the AFRCC, AKRCC's mission is derived from ICAO's Chicago Convention and the IAMSAR manual, along with the NSP, NSS and DoDD 3003.01, DoD support to civil SAR. Additionally, by agreement with USCG District 17, Juneau, Alaska, the AKRCC will assume SMC for aeronautical SAR events in Cook Inlet above 60 degrees north latitude, while the Juneau RCC will assume SMC for land SAR events below 58 degrees north latitude (Aleutian Chain) and east of 141 degrees west longitude (Southeast AK).

The AKRCC has no SAR command and control authority, but actively coordinates SAR operations of all available resources in Alaska. In addition to the normal DoD support to civil SAR on a non-interference basis with primary military duties common to all DoD assets, the Alaska ANG maintains a 24-hour alert response with HH-60G, HC-130 and Guardian Angel forces. In accordance with the SAR agreement between 11th AF and the Alaska ANG, mission priorities for the Alaska ANG rescue assets are:

1. RCC-directed SAR operations in support of DoD assets;
2. Combat mission readiness training;
3. RCC-directed civil SAR operations; and
4. Operational support to DoD.

In accordance with these priorities, alert force response posture is dictated by a combination of fighter aircraft operations and time of day.

AKRCC receives all Cospas-Sarsat distress beacon alerts located in the Alaska SRR. AKRCC shares distress beacon information with civil SAR agencies with SMC responsibilities and SAR responders. AKRCC works closely with the FAA to respond to audible aeronautical and ground distress beacon reports, investigates overdue aircraft reports, and determines appropriate type and scope of response to an incident.

AKRCC can be contacted at:

- Toll-Free: 1-800-420-7230
- Commercial: 907-551-7230
- DSN: 317-551-7230
- Fax (comm. or DSN prefix): 551-7245
- Email: rcc.messages@elmendorf.af.mil.

RCC Honolulu

USCG RCC Honolulu operates under Commander, 14th Coast Guard District in Honolulu, Hawaii.

RCC Honolulu has responsibility for:

- Maritime and aeronautical SAR within the U.S. SRR that includes a large portion of the Northern Pacific Ocean; and
- Aeronautical and other types of land SAR in the State of Hawaii, as requested (most land SAR operations in Hawaii are conducted by State and local SAR authorities).

As a USCG RCC, RCC Honolulu's specific SAR guidance is contained in the U.S. Coast Guard Addendum to the NSS.

National Park Service (NPS)

While the NPS does not operate a Federal RCC, it does, however, maintain a national

Emergency Incident Coordination Center (EICC) that is responsible for coordinating response resources throughout the NPS system.

National Parks are located throughout the U.S., with diverse physical environments and a diverse visiting public.

Varying levels of SAR services may be required based on the challenges associated within each Park. Park SAR operations are normally conducted by highly trained Park Rangers. Each Park has SAR plans that identify possible scenarios and response procedures.

As one of three Federal Agencies responsible for conducting SAR operations (the USCG and DoD are the other two), NPS works closely with Federal RCCs and other authorities and volunteers to provide mutual support.

Canada and Mexico

International SAR agreements have been concluded between the U.S. and several other countries.

SAR across the land border with Mexico is authorized through diplomatic channels. Maritime SAR cases are coordinated with the Mexican Navy.

Canada has three RCCs which cooperate closely with U.S. RCCs:

- JRCC Halifax;
- JRCC Trenton; and
- JRCC Victoria.

Canadian land SAR, similar to the situation in the U.S., is largely conducted under the legal authority of the individual Provinces and Territories. The Canadian Cospas-Sarsat MCC is co-located with Joint RCC (JRCC) Trenton.

National Park Service:
Emergency Incident Coordination Center

NPS operates an Emergency Incident Coordination Center (EICC) located in Luray, Virginia and is staffed and operated on a 24 hour basis, 365 days a year. Although each Park may have sufficient resources to accomplish SAR missions through both local staff and other State, local and volunteer SAR responders, the EICC may coordinate the movement and mobilization of supplemental personnel, or specialized resources including, aircraft, and small boats.

The EICC can be contacted at: (540) 999-3412.

This page intentionally left blank.

Section 1-5: Non-Federal SAR Responsibilities

Introduction

In accordance with the NSP, State, Tribal, Territorial, and local SAR authorities are responsible for land-based SAR and for designation of an SC within their respective jurisdictions. State SCs are integral partners with Federal SCs and are critical to providing effective civil SAR services.

State Governors have agreements with Federal SAR authorities concerning Federal-State SAR coordination and responsibilities. Most land SAR is conducted by local SAR authorities; however, when local and State capabilities are exceeded or special capabilities are required, Federal support may be requested via an RCC, local military command, or FEMA, as appropriate.

State SAR Responsibilities

The NSS discusses Federal and State SAR agencies, plans and resources, as well as Federal-State coordination and Federal assistance to States.

State responsibilities are normally documented in State legislation, plans, and/or agreements. Emergency response operations are normally organized and conducted in accordance with NIMS/ICS.

ICs are typically pre-designated local authorities, but could be a State or Federal person depending on the nature and magnitude of the incident.

Typically, a State Governor will assign responsibilities for emergency response to a State agency. In many States, a person is designated as State SC/Responsible Authority and performs RCC liaison and coordination. In many States, the State SC/Responsible Authority is also responsible for homeland security, public safety/law enforcement, emergency management, and aviation.

Each State, the District of Columbia, and the three U.S. Territories (Guam, Puerto Rico and the U.S. Virgin Islands) have Army National Guard (ARNG) and ANG resources that can be requested to support State SAR operations. These resources are under the direct control of the State Adjutant General.

More National Guard information is available at: http://www.ng.mil.

State SAR Plans

SAR agreements between a Federal Agency and a State should comply with relevant provisions of the NSP, NRF, the Federal Agency's applicable directives and applicable law.

Each State should have SAR plans that encompass civil (routine) and CISAR operations. The provisions of such plans vary depending on the State's requirements and accepted SAR responsibilities. Working relationships between a State and an RCC are established by agreement.

(Note: AFRCC maintains Federal-State SAR agreements, which can be obtained by contacting the AFRCC.)

States may have a variety of agreements that in some way support civil SAR; these could be with local, Tribal, or Territorial authorities, commercial entities, volunteer organizations, etc.

Emergency Management Assistance Compact (EMAC)

Every State is a member of the Emergency Management Assistance Compact (EMAC), a national interstate mutual aid agreement that enables States to share resources during times of disaster. EMAC has become the nation's system for providing mutual aid through operational procedures and protocols that have been validated during disasters where assistance was provided under the compact.

EMAC acts as a complement to the national disaster response system, providing relief to States requesting assistance from compact members struggling to preserve life, the economy, and the environment. EMAC does not replace Federal assistance, but can be used in conjunction with or when Federal assistance is not providing needed goods and services to an affected State. Requests for assistance are made at the discretion of the affected State; States retain the choice of seeking resource support from other States, the Federal Government or both, depending on the situation.

Further information on EMAC is available at: http://www.emacweb.org/ .

Tribal, Territorial, and Local SAR Responsibilities

Native American and Alaska Native communities vary in geographic size, complexity, and sophistication. Larger Tribes and communities have well established public safety infrastructure that includes law enforcement, emergency medical services, structural fire, and wild land fire capabilities. Each Tribe or community may elect to organize themselves with other local resources or maintain their own capability. In either case the Tribal Chief Executive has authority over actions and personnel working on Tribal lands.

DOI is charged with managing Federal affairs within U.S. Territories. DOI has a wide range of responsibilities (including the regulation of Tribal and Territorial governments, basic stewardship of public lands, etc.), but is not responsible for local government or for civil administration.

Tribes and Territories may request assistance directly from Federal Agencies. In cases where Tribes and Territories are seeking Federal assistance under the Stafford Act, these requests are submitted through ESF #15, External Affairs.

Section 1-6: International SAR System

Background

The international SAR system, used extensively worldwide since the 1950s, was first institutionalized under ICAO's Chicago Convention (Annex 12 – Search and Rescue) for international civil aviation, and later for maritime SAR by IMO, with associated guidance and standards provided in the IAMSAR Manual. Today, most nations use and supplement the international SAR system for national use as practicable for civil SAR. The U.S. has used the international SAR system for decades and continues to provide for its use via the NSP.

The Chicago Convention's Annex 12 applies to the establishment, maintenance, and operation of SAR services in the territories of Contracting States to the Convention, over the high seas, and for coordination of SAR operations between nations.

Annex 12 includes this note:

"This Annex is supplemented by the International Aeronautical and Maritime Search and Rescue (IAMSAR Manual) Manual, Volume I — Organization and Management, Volume II — Mission Co-ordination, and Volume III — Mobile Facilities (Doc 9731), the purpose of which is to assist States in meeting their search and rescue (SAR) needs and obligations accepted under the Convention on International Civil Aviation. These obligations, as they relate to the provision of SAR services, are specified in this Annex as Standards and Recommended

Practices. The three volumes of the IAMSAR Manual provide guidance for a common aviation and maritime approach to organizing and providing SAR services. States are encouraged, by use of the manual, to develop and improve their SAR services and to cooperate with neighboring States."

IMO's 1979 International Convention on Maritime Search and Rescue ("SAR Convention") provides that Parties follow relevant IMO guidelines for SAR, which are incorporated into the IAMSAR Manual.

The U.S. is signatory to both the Chicago and Maritime SAR Conventions.

Annex 12 and the SAR Convention are the basis for developing and implementing international SAR plans so that no matter where a distress situation occurs, persons in distress will be rescued by a SAR organization, and when appropriate, by cooperating SAR organizations of neighboring countries.

Parties to these conventions must ensure that effective arrangements are in place to provide adequate SAR services, and should enter into agreements with neighboring nations involving the establishment of SAR regions (SRRs), pooling of SAR facilities, establishment of common procedures, training, liaison, and the implementation of measures to expedite entry of SAR facilities into the territory of other Parties.

Plans developed under both conventions divide the globe into SRRs, each with an RCC, to help ensure the provision of adequate shore-based communication infrastructure, efficient distress alert routing, and proper operational coordination to support SAR services. Under this geographically-based scheme, RCCs respond to persons in distress without regard to the nationality or status of the persons or the circumstances in which they are found.

SAR Stages

As discussed in the IAMSAR Manual (Volume 2), the response to a SAR incident usually proceeds through a sequence of five stages, which are.

Associated with these stages are groups of activities typically performed by the SAR system in responding to a SAR incident from the time the system becomes aware of the incident until its response to the incident is concluded. The response to a particular SAR incident may not require performance of every stage. For some incidents, the stages may overlap with actions for two or more stages being performed simultaneously. The five SAR stages are described in Figure 1-6-1 on the next page.

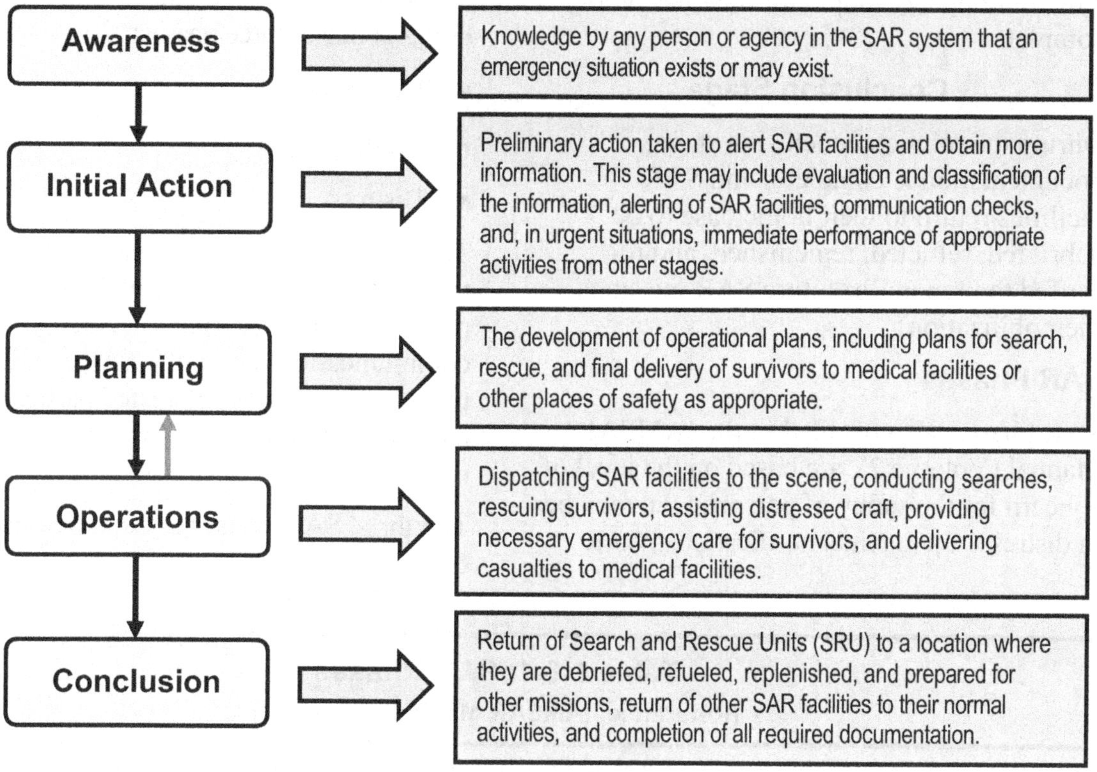

Figure 1-6-1: SAR Stages

Awareness Stage

During the awareness stage, gather and evaluate information to determine the nature of the distress, the appropriate emergency phase classification, and actions to take. Investigation should begin. Determine and record the information indicated below:

- Type of incident;
- Location and time;
- Search object description;
- Number and condition of people involved;
- Assistance desired;
- Weather on scene; and
- Reporting party information

Initial Action Stage

During the initial action stage, designate the SMC/IC, alert SAR facilities, set up an Incident Command Post (ICP) as appropriate, and evaluate the incident for urgency and extent of SAR response required.

Planning Stage

During the planning stage, develop operational plans for the incident-specific search, recovery, and delivery to a place of safety (search action plans and rescue action plans).

Operations Stage

During the operations stage, dispatch SAR resources to the scene, implement action plans developed during the planning stage, provide initial medical care, and deliver survivors to medical facilities or place of safety. If the search continues over several operational periods, the planning and operations stage will continuously loop until the mission is complete.

Conclusion Stage

During the mission conclusion stage, documentation is complete; and SAR facilities return to their home base to be debriefed, refueled, replenished, and re-staffed to prepare for future SAR missions (demobilization).

SAR Phases

SAR phases, also detailed in the IAMSAR Manual (Volume 2), are based on the level of concern for the safety of persons who may be in distress. Upon initial notification, a SAR incident is classified by the notified RCC as being in one of three phases:

- Uncertainty;
- Alert, or
- Distress.

The emergency phase may be reclassified by the SMC as the situation develops. The current phase should be referred to in all communications about the SAR incident as a means of informing interested parties of the current level of concern for the safety of persons who may need assistance.

The three SAR phases are detailed in Table 1-6-1 below:

Table 1-6-1: SAR Phases
(Reference: IAMSAR Manual, Volume 2)

Phase	Description
Uncertainty	Declared when there is doubt about the person(s) who may be in distress. The situation may need to be monitored and investigated, and more information gathered by a person trained in SAR investigative techniques. Additionally, preliminary communication (PRECOM) actions may begin (See Note below). This phase is generally investigative in nature with a low urgency in committing SAR facilities. *(Note: PRECOMs are preliminary communication searches; initial limited communications checks normally directed by the SMC during the uncertainty phase, of areas where the missing persons may be.)*
Alert	Exists when the person(s) in distress has difficulty and may need assistance, but is not in immediate danger or in need of immediate response. This phase may be indicated by a continued lack of information concerning location or progress when persons are overdue. PRECOM may be followed by Extended Communication (EXCOM) searches and SRUs might be dispatched to investigate high probability locations (See Note below). Other facilities might be alerted to the possibility of a mission. This corresponds to the intermediate or measured response score on the urgency determination for use of ground searches. *(Note: EXCOMs are extended communications searches to find information or clues about the location of missing persons; EXCOMs are normally conducted after PRECOMs have yielded no results or when the mission is upgraded to the alert phase.)*
Distress	Reasonable certainty that the craft or person requires an immediate response. May be indicated when all reasonable communications searches and other forms of investigation have not succeeded and no other actions are likely to be able to confirm non-distress. SRUs are launched immediately and/or other facilities are requested to take immediate action. This is the highest urgency or urgent response.

Incidents

An incident begins when the SAR system is notified of an actual or potential distress situation. An incident tracking number is assigned by the RCC and an SMC/IC is designated. Actions associated with alert and uncertainty phases are conducted, including incident evaluation, SAR resource alerting and PRECOM and EXCOM searches. An incident is an investigative function that does not normally involve dispatching SAR facilities to conduct SAR operations.

IAMSAR Manual (Volume 2) provides general considerations when evaluating and re-evaluating incident information:

To co-ordinate a response to a SAR incident most effectively, the SMC should have accurate, timely, and complete information about the incident and the subsequent status of survivors. Usually, not all the needed information is provided to the SMC. In fact, the fate of the distressed craft and any survivors is often a mystery in the early stages of the incident. For this reason, the SMC should initiate and actively pursue an investigation of the incident and related circumstances so the needed information can be obtained. Such investigative efforts are often similar in outline to scientific or police investigations. The SMC should interview, or have other qualified persons interview, anyone who might have some knowledge of the distress incident, distressed craft, or the persons aboard the distressed craft. These interviews may lead to additional persons, agencies, or other sources of information. The SMC should try to determine the most likely cause of the distress, if not already known, by consulting weather bureaus, ships, and aircraft, and by locating any known hazards to the distressed craft on an appropriate map or chart. There is almost no limit to the number of possible sources of information or the number of possible scenarios for what might have happened to the distressed craft. This means that the SMC should perform two apparently opposite activities: think of additional possibilities which should be investigated, and try to eliminate the maximum number of possibilities from further consideration through the investigative process.

If the search object is discovered during the investigation, the SMC/IC must deactivate all alerted resources involved in the incident.

Missions

A mission is a SAR operation conducted to assist persons who are in actual or apparent distress. Such persons may be aboard an aircraft, vessel or other craft. A mission is initiated when there is reasonable certainty that persons are in distress and may need immediate assistance, or if the existence of a distress situation cannot be determined by other means.

A mission number is assigned and the incident SMC/IC or another SMC/IC is designated for the mission.

Before issuing a mission number, the SMC/IC should determine that:

- A valid or perceived distress situation exists (normally assessed in terms of a threat to life, limb, or eyesight or of undue suffering);
- The responsible agency has requested the mission; and
- No known legal restrictions on activities or funding apply (e.g., Posse Comitatus Act, Stafford Act, etc.).

Missions normally fall into one of the mission types identified in Table 1-6-2 on the following page:

Table 1-6-2: Land SAR Missions	
Aircraft Missions	• Aircraft - search involving overdue GA, DoD, and commercial aircraft; • Precautionary - pre-positioning of SAR forces for use during a scheduled activity where the potential for SAR is high (e.g. in-flight emergencies, air shows, etc.).
Distress Beacon Missions	• *Emergency Locator Transmitters (ELTs)* - emergency distress beacons that most U.S. civil aircraft are required to carry; these distress beacons alert via satellites on 406 MHz or via audibly to overflying aircraft or ground stations on 121.5 MHz; • *Emergency position-indicating radio beacons (EPIRBs)* – emergency distress beacons that many ships, boats, and other marine craft carry that alert via satellites on 406 MHz; and • *Personal locator beacons (PLBs)* – multi-environment emergency distress beacons that individuals carry that alert via satellites on 406 MHz. Each of these types of distress beacons typically transmits a continuous 121.5 MHz signal that SAR forces can use for homing, and SAR forces may also be equipped to home on the 406 MHz alerting signal. Any of these beacons can be activated manually, but many ELTs and EPIRBs are designed to activate automatically when an aircraft crashes or a vessel sinks.
Non-Aircraft Missions	Support to recognized State or other SAR authorities for missions not involving aeronautical SAR such as missing person(s), rescue, medical evacuation, patient transport, mercy missions (e.g. transport of organs, blood, and specialized equipment), mass rescue operations (MROs), or transport of SAR personnel or facilities.

Concluding a SAR mission

When the search object is found, the SAR case is closed. When the search object is not located, the SAR mission is eventually suspended or terminated. Concerning the suspension or termination of SAR operations, the NSP states the following:

National SAR Plan: Suspension or Termination of Operations

Civil SAR operations shall normally continue until all reasonable hope of rescuing survivors has passed.

The responsible RCC/RSC concerned shall normally decide when to discontinue these operations. If no such center is involved in coordinating the operations, the On Scene Commander (OSC) or Incident Commander (IC) may make this decision. If there is no OSC or IC involved, the decision will be made at an appropriate level of the chain-of-command of the facility conducting the operations.

When an RCC/RSC or other appropriate authority considers, on the basis of reliable information that a rescue operation has been successful, or that the emergency no longer exists, it shall terminate the SAR operation and promptly so inform any authority, facility or service which has been activated or notified.

If an operation on-scene becomes impracticable and the RCC/RSC or other appropriate authority concludes that survivors might still be alive, it may temporarily suspend the on-scene activities pending further developments, and shall promptly so inform any authority, facility or service which has been activated or notified. Information subsequently received shall be evaluated and operations resumed when justified on the basis of such information.

Normally, the SMC is best qualified to decide or to recommend to an IC or higher authority when to suspend or terminate a mission. The IC or higher authority within the SMC's agency may reserve the right to make the final determination, taking into account the SMC's recommendation.

Closed, suspended, or terminated missions may be classified as per Table 1-6-3 below:

Table 1-6-3: Classification of Closed, Suspended, or Terminated SAR Missions	
Self-Recovered	Person(s) was reported lost or missing, but returned without assistance, or did not request further assistance
Saved	Survivor(s) was in distress, could not self-recover, and was rescued;
Deceased	Person(s) who is either known to have died, determined to have died on the basis of conclusive evidence, or declared to be dead on the basis of a presumptive finding of death. The recovery of remains is not a prerequisite to determining or declaring a person deceased.
Missing	Person(s) remains unlocated after all leads and search efforts have been exhausted.

Mission Suspension

When a SAR case cannot be closed and further search efforts appear futile, the search may be suspended, but only when there is no longer any reasonable hope of rescuing survivors from the SAR incident. Considerations for suspending a search include:

- All potential areas have been thoroughly searched;
- All reasonable probable locations have been investigated;
- All reasonable means of obtaining information about the whereabouts of the search object have been exhausted; and
- All assumptions and calculations used in search planning have been reviewed.

The SAR case will remain open (not closed/terminated) until the object of the search is located. If new information is received indicating the object of the search may not have been in the areas searched, or pertinent details of the search object were other than those previously assumed or reported, the suspended search may be resumed.

The decision to suspend a mission is a judgment call based on a careful and experienced analysis of the factors of a particular case.

If an aircraft is not located, the mission is not closed, but remains suspended until the aircraft is located or confirmed to have crashed.

Mission Termination

SAR operations are normally continued until all reasonable hope of rescuing survivors has passed; this determination is normally best made by the SMC/IC.

The SMC/IC will normally decide when to terminate SAR operations, typically not until all crews return to their place of origin.

Distress beacon missions are terminated once the signal source is either located, or the signal ceases without any additional clues to further an investigation.

For Federal SMCs/ICs, non-aircraft SAR missions are terminated when the search object is:

- Located and delivered to a medical facility or place of safety;
- Determined to be criminal activity or threat to SAR resources; or
- It is verified that a distress situation no longer exists.

Aircraft missions are terminated once the aircraft is located or contact has been made with the pilot to confirm that no distress exists.

Section 1-7: NIMS

Incident Command System Overview

HSPD-5

NIMS – Six Components

Incident Command System (ICS) Overview

ICS, now a substantial part of NIMS, is designed to provide a commonly accepted management structure for better decisions and safer, more effective use of available resources to achieve tactical objectives for incidents that involved many Federal, State, Tribal, Territorial, and local agencies. ICS provides a flexible structure that can be adjusted to incident demands with reduced concerns about jurisdictional boundaries.

In 1994, NSARC formally requested FEMA to become the national sponsor of ICS. FEMA subsequently delegated much of this responsibility to the National Fire Academy, which now leads in ICS standards and training activities.

HSPD-5

In 2003, Homeland Security Presidential Directive 5: Management of Domestic Incidents (HSPD-5) identified steps for improved coordination in response to incidents and required the Department of Homeland Security (DHS) to coordinate with other Federal Departments and Agencies and with local, State, Tribal, and Territorial governments to establish a National Response Plan (NRP) and a National Incident Management System (NIMS). The NRP was subsequently replaced by the National Response Framework (NRF).

The NRF is an all-discipline, all-hazards plan for the management of domestic incidents using the template established by NIMS to coordinate and integrate incident management and emergency support functions across local, State, Tribal, Territorial, and Federal government entities, the private sector, and nongovernmental organizations.

NIMS – Six Components

NIMS and ICS are not the same; ICS is one facet of NIMS. NIMS is an approach to domestic incident management for all jurisdictional levels and across functional disciplines in an all-hazards context. Six major components make up the NIMS approach:

- *Command and Management* - NIMS standard incident command structures are based on three key organizational systems:
    - ICS defines the operating characteristics, interactive management components, and structure of incident management and emergency response organizations engaged throughout an incident;
    - Multiagency Coordination Systems define the operating characteristics, interactive management components, and organizational structure of supporting incident management entities engaged at the Federal, State, local, tribal, and regional levels through mutual-aid agreements and other assistance arrangements; and

- Public Information Systems refers to processes, procedures, and systems for communicating timely and accurate information to the public during crisis or emergency situations;

- *Preparedness* - Effective incident management begins with activities conducted on an ongoing basis in advance of any potential incident. Preparedness involves integrated planning, training, exercises, personnel qualification and certification standards, equipment acquisition and certification standards, and publication management processes and activities;
- *Resource Management* - NIMS defines standardized mechanisms and establishes requirements for processes to describe, inventory, mobilize, dispatch, track, and recover resources over the life cycle of an incident;
- *Communications and Information Management* - NIMS requires a standardized framework for communications, information management (collection, analysis, and dissemination), and information sharing at all levels of incident management;
- *Supporting Technologies* - Technology and technological systems support implementation and refining NIMS. These technologies include communications systems, information management systems, data display systems and specialized technologies; and
- *Ongoing Management and Maintenance* - This involves NIMS strategic direction and oversight.

For ICS, an incident is an occurrence caused by people or natural phenomena that requires response to prevent or minimize loss of life or environmental or property damage. Examples of incidents include: structural and wild land fires; natural disasters; disease outbreaks; search and rescue; hazardous materials incidents; terrorist and criminal acts; and security events such as Presidential visits or the Super Bowl.

Section 1-8: International SAR System - NIMS/ICS System Integration

Introduction

NIMS/ICS and the international SAR system are both used for SAR, especially land SAR, and may be used separately or together.

NIMS/ICS provides for a consistent approach for Federal, State, Tribal, Territorial, and local governments, the private-sector and nongovernmental organizations to work effectively and efficiently together to prepare for, respond to, and recover from domestic incidents, regardless of cause, size or complexity. In the use of NIMS/ICS, the event may involve significant and complex challenges, where SAR may be a significant portion or only have a limited role. Normally, the use of NIMS/ICS is incident specific, being implemented when required for coordinating interagency resources for a specific event.

The international SAR system has been used in the U.S. and worldwide for decades in the conduct of aeronautical and maritime SAR. Every year, tens of thousands of SAR missions are conducted by Federal Agencies in the U.S. aeronautical and maritime SRRs using the international SAR system. Normally SAR missions using the international SAR system are coordinated by an RCC responsible for a specific aeronautical or maritime SRR.

The international SAR system is used exclusively for civil SAR, while NIMS/ICS is used for civil SAR and other types of emergency response operations.

The U.S. SAR system is based on the NSP, which is in part based on various international treaties, conventions, and agreements.

NIMS, and NIMS-based SAR operations on the other hand, are implemented as directed by Homeland Security Presidential Directive (HSPD) – 5 (Management of Domestic Incidents).

The international SAR system and NIMS are both essential and are mutually compatible.

The international SAR system accommodates standard international SAR organizational structure, procedures, and terminology, especially as used for aeronautical and maritime SAR.

The international SAR system and NIMS are used in combination for CISAR. CISAR consists of civil SAR operations carried out as all or part of the response to an emergency or disaster declared by the President under provisions of the Stafford Act, the NRF, and ESF #9.

Understanding certain terms (see the Glossary) helps to understand the two systems. Table 1-8-1 provides a comparison between both the international SAR system and NIMS/ICS.

Table 1-8-1: International SAR System-NIMS/ICS: Terminology Comparison	
International SAR System Terms	**NIMS/ICS Terms**
Catastrophic Incident SAR (CISAR)	
Civil search and rescue (SAR)	Area Command (AC)
Mass Rescue Operations (MROs)	Emergency Management Assistance Compact (EMAC)
National Search and Rescue Committee (NSARC)	Emergency Support Function (ESF)
National Search and Rescue Plan (NSP)	Incident Command Post (ICP)
On-Scene Coordinator (OSC)	Incident Command System (ICS)
Rescue Coordination Center (RCC)	Incident Commander (IC)
SAR Coordinator (SC)	National Incident Management System (NIMS)
SAR facility	National Response Framework (NRF)
SAR Mission Coordinator (SMC)	Operations Section Chief (OSC)
SAR Region (SRR)	Primary Agency
SAR service	Unified Command (UC)
SAR Unit (SRU)	

Incident Command System (ICS)

ICS is used by both the international SAR system and NIMS. The difference is that the international SAR system uses ICS, as warranted, for supplementing standard coordination procedures used worldwide for SAR, while use of ICS for operations covered by NIMS is standard.

NIMS uses ICS exclusively for incident management and is an integral part of NIMS. ICS:

- Has a flexible, scalable, and adaptable coordinating structure (Figure 1-8-1);
- Aligns key roles and responsibilities across jurisdictions;
- Links all levels of government, private sector, and nongovernmental organizations in a unified approach to emergency management; and
- Is always in effect and can be partially or fully implemented.

The centerpiece of ICS is a standard organizational structure to aid incident management activities (Figure 1-8-1).

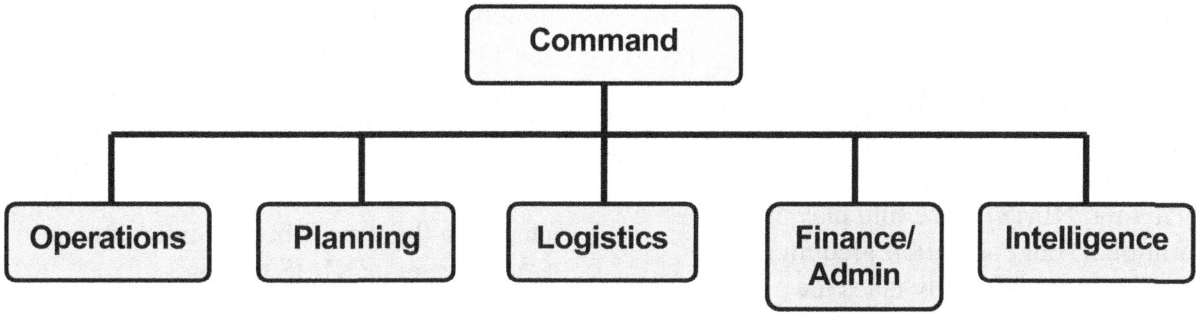

Operations	Planning	Logistics	Finance/ Administration	Intelligence
Conducts tactical operations to carry out the plan. Develops tactical objectives and directs all resources analysis.	Develops the action plan to accomplish the objectives; collects and evaluates information; maintains resource status.	Provides support to meet incident needs: provides resources and services.	Monitors incident costs; provides accounting, procurement, time recording, and cost analysis.	Provides analysis and sharing of information and intelligence during an incident.

Figure 1-8-1: Basic ICS Structure

International SAR System - NIMS/ICS Comparison

Appendix A compares key aspects of the SAR System and NIMS.

Some important facts to note include:

- The international SAR system must be used for certain types of SAR, and can be invaluable to use for other types, but most emergency responders, especially for land SAR, rarely need to know about or use it (a sample exception would be an SMC/IC handling a major or international aeronautical or maritime SAR case);
- The international SAR system is used worldwide, while NIMS is used less outside the U.S.;
- The international SAR system is specialized for SAR, while NIMS is designed for any emergency;
- An SMC working in an RCC is normally responsible for coordinating SAR response under the international SAR system, while an IC (with or without the support of an SMC) working in an ICP would commonly be responsible when NIMS is in use;
- Both the international SAR system and NIMS can accommodate distress incidents of any size;
- The international SAR system uses ICS to supplement international procedures as appropriate, whereas NIMS uses ICS consistently along with the international SAR system as appropriate; and
- The international SAR system and ICS are mutually compatible.

Combining the International SAR system and NIMS/ICS

Sometimes SAR responders not familiar with both the international SAR system and NIMS conclude that one or the other should be abandoned. However, for both legal and practical reasons this cannot and should not be done. The international SAR system and

NIMS are fully compatible and work well in combination.

For example, when a disaster is declared by the President and CISAR operations are conducted under the NRF's ESF #9, the NSP still applies, and the provisions of the NRF (and NIMS) come into play as additional (but compatible) requirements. This process essentially takes the international SAR system and plugs it into NIMS by linking it with one or more appropriate SAR representatives supporting the IC in the Operations Section of the Incident Command.

SAR personnel can perform functions such as supporting the IC, serving as the IC's liaison with the international SAR system, helping to coordinate use of resources for the overall response (typically with priority given to lifesaving), and requesting support for SAR from the IC without significantly changing the way either the international SAR system or NIMS operates.

Often one of these SAR personnel can serve as the SAR Branch Director or SAR Group Supervisor for the IC within the command structure.

Section 1-9: Special SAR Operations

Mass Rescue Operations (MROs)

Catastrophic Incident Search and Rescue (CISAR)

The "Olive" SAR Model

Mass Rescue Operations (MROs)

Mass Rescue Operations (MROs) are internationally characterized by the need for immediate response to large numbers of persons in distress, such that the capabilities normally available to SAR authorities are inadequate. What constitutes "large numbers of persons" for this definition will vary based on factors such as weather, location, available SAR resources, type of scenario, and condition of the survivors.

MROs are low probability, high consequence events that occur less frequently than typical, or routine, SAR operations. MROs pose special challenges for SAR planners and responders.

MROs often must be carried out and coordinated within a broader emergency response context that may involve hazards mitigation, damage control and salvage operations, pollution control, complex traffic management, large scale logistics, medical and coroner functions, accident-incident investigation, intense public and political attention, etc. The priority of the MRO response should always be lifesaving, followed by other efforts such as environmental protection, saving property, etc.

Careful and comprehensive planning, preparation, and training are essential for effectively responding to an MRO.

Catastrophic Incident Search and Rescue (CISAR)

Catastrophic Incident

"A catastrophic incident is any natural or manmade incident, including terrorism, which results in extraordinary levels of mass causalities, damage, or disruption severely affecting the population, infrastructure, environment, economy, national morale, and/or government functions."

National Response Framework, page 42

The CISAR Addendum provides guidance for SAR operations during Catastrophic Incidents.

CISAR consists of civil SAR operations carried out as all or part of the response to an emergency or disaster declared by the President, under provisions of the NRF and Emergency Support Function (ESF) #9.

(Note: Normally a Temporary Flight Restriction (TFR) will be implemented over the affected area. Consideration must be given to the dimensions of the TFR as well as the entry requirements and procedures when operating within the TFR.)

The nature of CISAR could range from normal SAR operations to the conduct of MRO. Two criteria must be met for an incident to be identified as CISAR. First, the response must be associated with a Presidential Declaration. Second, ESF #9 must be implemented. Clear delineation between normal SAR, MRO, and CISAR may not be apparent. However, it is important to understand that flexible response options are available for these progressive or potentially overwhelming events. The CISAR Addendum provides standardized guidance for the conduct of CISAR operations.

Provisions of the NSP, NSS, and relevant addenda always apply to civil SAR regardless of whether the operations are CISAR. If the operations are CISAR, then provisions of the NRF and its relevant supporting documents also apply.

For most CISAR operations, the NRF's ESF #9 (Search and Rescue) Annex will be activated and an overall Primary Agency for the event will be designated by the ESF #9 Coordinator (FEMA). The ESF #9 overall Primary Agency designation is dependent on the type of catastrophic incident (Structural Collapse SAR – DHS/FEMA; Maritime/Coastal/Waterborne SAR – DHS/USCG; or Land SAR – either DoD or DOI/NPS) and the nature of the SAR response.

Another important aspect of CISAR is that Federal Agency costs for such operations are generally reimbursed under provisions of the Stafford Act. For example, for NRF related operations in support of ESF #9, DoD's role in CISAR is always considered to be a DSCA event and is conducted in accordance with DoD's DSCA directives. Most Federal Agencies have similar policy documents relating to Stafford Act operations.

CISAR operations are covered in the CISAR Addendum to the NSS.

The "Olive" SAR Model

NSARC created a SAR model that categorizes the severity of SAR operations conducted by the NSARC member agencies and other SAR authorities. This SAR model has been called the "Olive" (Figure 1-9-1).

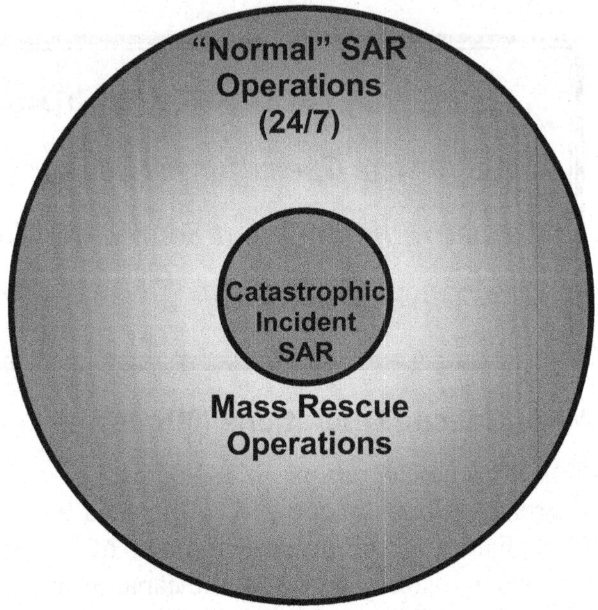

Figure 1-9-1: The "Olive"

In the Olive, the green area represents SAR operations conducted under the authorities of the NSP by designated Federal SCs and other SAR authorities on a daily basis (e.g., NPS, U.S. Coast Guard, and CDRUSNORTHCOM/ CDRUSPACOM)

The yellow area represents MROs; SAR operations not considered "normal" or routine SAR, but also do not meet the criteria of a Catastrophic Incident.

The red area represents civil SAR operations when a catastrophic incident has occurred and will normally be associated with a Presidential emergency or disaster declaration and additional provisions of the NRF (ESF #9) will apply. As stated in the previous section, the NSP still applies, and the provisions of the NRF (and NIMS) come into play as additional (but compatible) requirements.

There is no line between "normal" SAR operations and MROs. Each event is unique to each agency, circumstances surrounding the incident, SAR resources available to respond, type of SAR conducted (e.g., land, aeronautical, maritime, urban, etc.), weather and on scene conditions, as well as many other factors. There is a line between catastrophic incidents and MROs to symbolize that a catastrophic incident is a clearly defined NRF incident normally associated with a Presidential declaration.

The purpose of the Olive is to highlight the differences and understand the challenges associated with each type of SAR response. NSARC member agencies, States, Tribes, Territories, and local SAR authorities need to plan for regular SAR, but also for possible MROs and CISAR as well.

This page intentionally left blank.

Part 2: Resources and Capabilities

This page intentionally left blank.

Section 2-1: Land SAR Resources

Introduction

Aviation Resources

Rotary Wing Aircraft

Fixed Wing Aircraft

Land Resources

Ground Searchers

Search Dogs

Human Trackers

Rough Terrain SAR Responders

Swift Water and Underwater SAR Responders

Winter Environment SAR Responders

Specialized Vehicle SAR Responders

Confined Space SAR Responders

Collapsed Structure SAR Responders

Horse/Mounted SAR Responders

Department of Defense (DoD)

84th Radar Evaluation Squadron (RADES)

National Park Service (NPS)

National US&R Response System

U.S. Coast Guard (USCG)

Volunteer Auxiliaries

Civil Air Patrol (CAP)

Coast Guard Auxiliary (CGAUX)

Section 2-1: Land SAR Resources (continued)

Federal Aviation Administration

Air Traffic Organization (ATO)

Flight-Following and Alerting Services

Radar Nets

National Track Analysis Program (NTAP)/National Track Analysis (NTA)

National Search and Rescue School

Nationally-organized Resources

National Association for Search and Rescue (NASAR)

Mountain Rescue Association (MRA)

National Cave Rescue Commission (NCRC)

National Ski Patrol

National Voluntary Organization Active in Disasters (NVOAD)

U.S. SARSAT Program (Overview)

Cospas-Sarsat System Satellites

LEOLUTs and GEOLUTs

USMCC

Introduction

SMC/ICs and other SAR authorities involved in coordinating SAR operations need to:

- Identify the SAR resources available in their area;
- Coordinate their use to provide the appropriate and capable responses to a SAR mission; and
- Effectively employ SAR resources to their best advantage.

It is important that SMCs/ICs and SAR planners be familiar with the operational capabilities of aviation, land, and water based SAR resources in their AOR.

The following is a sampling of various types of SAR resources that may be utilized during a SAR case.

Aviation Resources

Rotary Wing Aircraft

Helicopters make excellent SAR platforms and are routinely used in land SAR operations for:

- Visual and electronic searching over priority search areas;
- Directing ground personnel to an area from the air;
- Inserting and/or extracting rescue personnel and victims; and
- Movement or transport of equipment and personnel.

Fixed Wing Aircraft

Fixed wing aircraft can be used for land SAR operations for:

- Visual and electronic searching large areas quickly;
- Searching for larger search objects, such as aircraft or automobiles; and
- Movement or transport of equipment and personnel.

Land SAR Resources

Ground Searchers

Ground searchers are used to conduct confinement, attraction, rapid (hasty) and area searches (Section 4-4). Search thoroughness largely depends on the type of terrain being searched, the size and appearance of the search object, the local environmental conditions, and the searchers' training and experience in clue detection.

Ground searchers:

- Provide rapid route or area search of a specific area;
- Are normally self sufficient and have good navigation skills; and
- Can locate clues to eliminate or redefine a search to a smaller search area.

Search Dogs

There are several canine SAR disciplines. As per Table 2-1-1, dogs can be classified broadly as either:

- Air scenting dogs; or
- Tracking/trailing dogs.

Biloxi, Miss., September 3, 2005 - A member of the Indiana Task Force 1 Urban Search and Rescue (US&R) team and his canine partner enter a damaged house to search for victims of Hurricane Katrina. (Photo: FEMA/Mark Wolfe)

Table 2-1-1: Search Canines – General Categories		
Type	**Description**	**Uses**
Air Scent Dogs	Air scent dogs primarily use airborne human, normally non-discriminate, scent to home in on subjects.	Air scent dogs can: • Search large areas efficiently; • Search difficult areas more easily than humans; and • Can easily integrate into a search area.
Tracking/ Dogs	Tracking dogs are scent discriminating, work on-lead and follow the footsteps of the subject utilizing ground scent.	• Tracking dogs are highly efficient; • Follow the ground scent of the last person who passed through the search area; • Are commonly used among law enforcement and correctional departments; • Discriminate a lost subject from other persons; and • Can function independently for long periods of time. (Note: Scent-article is not required.)
Trailing Dogs	Trailing dogs are scent discriminating, work on-lead or off-lead and follow the subject's path utilizing both air and ground scent.	• Trailing dog is scent-discriminating; • Follows the scent trail of a specific individual; • A scent article required; and • Handlers prefer to obtain scent-articles themselves.

Tracking dogs not only follow non-human ground disturbance scent, but also follow an individual's contact scent.

The main difference between a trailing dog and a tracking dog is the tracking dog works footprint to footprint with nose on the ground, while the trailing dog normally holds its head higher, does not work footstep to footstep and is usually farther from the actual track.

Both trailing and tracking dogs require a source of the individual's scent, but not necessarily an actual scent article. Both types of dog can be started off a footprint or contact point, from a gauze pad that has been allowed to absorb scent off of a car seat, steering wheel, etc., or a scent article such as a piece of the search object's clothing.

Other non-scent discriminate canine SAR disciplines are described in Table 2-1-2:

Table 2-1-2: Search Dogs – Non-Scent Discriminate Canine SAR Disciplines	
Type	**Description**
Wilderness Area	Used to search large areas for the presence of a live and/or human subject(s)
Cadaver/Human Remains Detection (HRD)	Used to search for the presence of human remains on the surface, hanging, or buried. They are used in a large variety of search environments, especially in disaster situations.
Water	Used from a boat or the shore to search for the presence of submerged human remains in water (salt or freshwater).
Avalanche	Used to search avalanche debris flows for the presence of humans buried in snow.
Disaster/US&R	Used to search debris piles or urban areas for the presence of live human subjects.

Human Trackers

Human trackers are searchers trained to follow the presumed route of the subject from the LKP/Point Last Scene (PLS) by locating tracks, sign, and other disturbances left by the search object. This technique is often referred to as "man-tracking" or just "tracking."

These specialized searchers:

- Eliminate large portions of search area;
- Quickly determine the subject's initial direction of travel; and
- Are highly efficient if a good LKP/PLS exists.

Rough Terrain SAR Responders

Rough terrain SAR responders are searchers that have technical skills and equipment to operate in a hilly, mountainous, urban, or a very remote environment. These specialized search personnel are usually members of:

- Local rescue or SAR teams;
- Fire departments; and/or
- Mountain Rescue Association (MRA).

Swift Water and Underwater SAR Responders

Swift water and underwater SAR responders are searchers that have technical skills and equipment to operate in surf, swift water, or deep-water conditions. These specialized search personnel are usually members of:

- Local rescue or SAR teams;
- Dive- rescue units; and/or
- Organized divers' organizations.

Winter Environment SAR Responders

Winter environment SAR responders are searchers that have technical skills and equipment to operate in snow, ice, avalanches, etc. These specialized search personnel are usually members of:

- Local rescue or SAR teams;
- National Ski Patrol;
- Ski area pro-patrols; and
- MRA.

Specialized Vehicle SAR Responders

Specialized vehicle responders are searchers or local units with vehicles that operate in special terrain or difficult environmental conditions. These specialized search personnel are usually members of:

- Local Rescue or SAR teams;
- Snowmobile or ATV associations;
- Local fire departments; and/or

- Mountain biking teams.

Confined Space SAR Responders

Confined space responders are searchers that have technical skills and equipment to operate in caves, caverns, mines, tunnels, shafts, tanks, etc. These specialized search personnel are usually members of:

- National Cave Rescue Commission/ National Speleological Commission;
- Fire services;
- FEMA;
- Mine Safety and Health Administration;
- Office of Surface Mining; and
- "Cavers" or local Speleological Groups.

Collapsed Structure SAR Responders

Collapsed structure responders are searchers that have technical skills and equipment to operate in collapsed structures.

Port-au-Prince, Haiti, January 18, 2010 – FEMA Urban Search and Rescue. Rescue of earthquake victims in collapsed structures. (Photo: USAID/Gruber)

These specialized search personnel are usually members of:

- Search dog units;
- Urban Rescue Teams; and
- FEMA US&R Task Forces.

Horse/Mounted SAR Responders

Horse/mounted SAR responders are trained searchers mounted on horses, with riding or pack animals used for searching remote or rough areas, or responders that assist by transporting supplies and equipment with horses and pack animals. Horse/mounted search personnel are usually members of:

- Organized mounted SAR units;
- Sheriff or auxiliary units;
- Local riding clubs or associations; and
- Ranchers, outfitters, wilderness guides, etc.

Department of Defense (DoD)

DoD civil SAR activities involve a broad spectrum of all domain resources. Depending on the scale of an incident, effective SAR operations may require any or all of the following:

- Incident, Awareness, and Assessment (IAA) capability (manned, unmanned, satellite);
- Light, medium, and heavy lift helicopters;
- Fixed wing aircraft;
- Small boat and utility vessels;
- Amphibious and high water vehicles;
- Search and extraction elements;
- K9 and general purpose personnel (Area search & US&R support); and
- Aircraft radar evaluation/tracking.

SAR planners need to understand that not all DoD resources are SAR-equipped and not all crews are SAR trained and experienced.

84th Radar Evaluation Squadron (RADES)

The 84th Radar Evaluation Squadron's (RADES) mission is to provide worldwide radar-centric planning, optimization, and constant evaluation to create the most sensitive integrated radar picture. Its Event Analysis (EA) process is a post-event investigation of aerial mishaps that has provided a unique and sometimes crucial perspective on a large number of military and civilian aircraft incidents. They regularly provide assistance to the AFRCC, Air Force Safety Center, Navy Safety Center, and NTSB.

The Radar Data Interface System (RDIS), designed and built by 84 RADES engineers and present at all Air Defense Sectors and Alaska, supports their EA process.

National Park Service (NPS)

NPS has a wide array of SAR capabilities and is equipped to deal with most emergencies that could arise in national parks (391 areas of 84 million acres) and monuments, as well as other SAR operations as requested outside the national parks.

National Park Service Rangers practicing water rescue (Photo: NPS)

NPS has developed several SAR qualifications for park rangers. These qualifications and the appropriate tasks are identified in Position Task Books (PTBs) in accordance with NIMS.

Each PTB lists the performance requirements (tasks) for a specific position in a format that allows a trainee to be evaluated against written guidelines. Successful performance of all tasks, as observed and recorded by an evaluator, will result in a recommendation that the trainee be certified in that position. A more detailed description of this process, definitions of terms, and responsibilities are included in the Wildland Fire Qualification System Guide (NWCG PMS 310-1) and National Park Service All-Hazard Incident Qualification System Guide.

NPS SAR position tasks include the following:

- *Mountain Search and/or Rescue Technician* searches and rescues those in trouble in urban, wilderness, mountainous environments, or environments usually greater than 25 degrees inclination, or wherever rope systems or three points of contact are commonly necessary, and includes areas commonly known as snow or ice covered terrain, glaciers, crevasse, backcountry, and alpine;

- *Swiftwater/Flood Rescue Technician* carries out rescues from shore and in water;
- *Wilderness Search and Rescue Leader* leads a SAR team or crew, searching for and rescuing those in trouble in wilderness, rural, urban, suburban and other environments; and
- *Wilderness Search and Rescue Technician* searches for and rescues those in trouble in wilderness, rural, urban, suburban and other environments.

National US&R Response System

FEMA, with support from Federal, State, Tribal, Territorial and local SAR authorities, the nation's top technical US&R specialists, and other interested groups, developed the National Urban Search and Rescue Response System. This system includes building/structural collapse SAR operations for natural disasters as well as other building collapse operations that primarily require DHS/FEMA US&R task force operations. The National US&R Response System integrates DHS/FEMA US&R task forces, Incident Support Teams (ISTs), and technical specialists.

The System is built around a core of task forces prepared to deploy immediately and initiate operations in support of ESF #9. These task forces are staffed primarily by local fire department and emergency services personnel who are highly trained and experienced in collapsed structure SAR operations and possess specialized expertise and equipment.

ISTs provide coordination and logistical support to US&R task forces during emergency operations. They also conduct needs assessments and provide technical advice and assistance to State, tribal, and local government emergency managers. The ISTs are staffed by personnel from US&R task forces; Federal, State, tribal, and local government emergency response organizations; and private-sector organizations.

Technical specialists provide expertise in various US&R disciplines and are mobilized as needed.

Sabine Pass, TX, September 14, 2008 - Members of the FEMA Urban Search and Rescue team, Indiana Task Force 1 go into neighborhoods impacted by Hurricane Ike to search for people needing help evacuating the area. (Photo: Jocelyn Augustino/FEMA)

U.S. Coast Guard (USCG)

The USCG maintains SAR resources, (aircraft, cutters, and boats) primarily dedicated to maritime SAR throughout the U.S. Because of the USCG's 24/7 readiness posture and the seamless facilities along the entire coast, Great Lakes, and inland waterways, the USCG is often available to support inland SAR.

USCG aircraft include long-range HC-130, medium-range HU-25 fixed-wing aircraft, HH-60and HH/MH-65 Rescue Swimmer/ EMT equipped helicopters that operate in any weather, day or night, with capabilities that include raft, and Datum Marker Buoy

(DMB) insertion, hoisting and sling operations, shipboard capability forward-looking-infrared (FLIR), and a wide spectrum of communications capability (VHF-FM, VHF-AM, UHF, HF and SATCOM) that may be utilized for IAA, search, rescue, and transport.

Coast Guard HH-65 Helicopter (Photo: Coast Guard)

USCG cutters include high-endurance cutters (WHECs), medium-endurance cutters (WMECs), patrol boats (WPBs), buoy tenders, icebreakers, and harbor tugs.

In addition, USCG Rigid Hull Inflatable Boats (RHIB), small boats, ice skiffs, and Special Purpose Craft (SPC) ranging up to 65 feet in length, are excellent resources for inland ice/waterborne SAR.

Volunteer Auxiliaries

The USCG Auxiliary (CGAUX) and the Civil Air Patrol (CAP) are volunteer civilian organizations that are auxiliaries to the USCG and USAF, respectively, and which have capabilities to support SAR missions.

Civil Air Patrol (CAP)

CAP is a congressionally chartered, nonprofit organization of experienced aviation-trained volunteers located in every State and Puerto Rico that also serves as a USAF auxiliary.

CAP operates 550 light civil aircraft that have land-based SAR and other capabilities including specialized sensors for direction finding and airborne digital imaging. A small number of strategically located aircraft operate a hyper spectral enhanced reconnaissance imaging system. CAP operates a nationwide communications system in support of its air and ground operations. CAP averages 80 to 100 saved lives per year and flies 90 percent of the land SAR missions approved by the AFRCC.

Further information on the CAP is available at: http://www.cap.gov.

Coast Guard Auxiliary (CGAUX)

Established by Congress, the CGAUX mission is to contribute to the safety and security of persons, ports, waterways and coastal region; the CGAUX balances the missions of recreational boating safety with homeland security.

As with all USCG resources, requests for and tasking of CGAUX resources are conducted through the appropriate USCG RCC.

Further information on the CGAUX is available at: http://nws.cgaux.org/ .

Federal Aviation Administration

The Federal Aviation Administration (FAA), operating under the Department of Transportation (DOT), provides air traffic control services to include monitoring and flight following aircraft through its Air Route Traffic Control Centers (ARTCC), Airport Traffic Control Towers (ATCT)/Terminal Radar Approach Controls (TRACON), and Flight Service Stations (FSS). The FAA may be the first agency to alert an RCC of an aviation emergency or overdue aircraft.

The FAA also monitors emergency frequencies for 121.5/243.0 MHz emergency locator transmitter (ELT) signals. Most aircraft are required by the FAA to carry ELTs that are designed to automatically activate in the event of a crash.

Further detailed information about the FAA and SAR (see FAA Order JO 7110.65 Air Traffic Control, JO 7110.10, Flight Services, or the Aeronautical Information Manual) are available at:

http://www.faa.gov/documentlibrary/media/order/atc.pdf

http://www/faa.gov/documentlibrary/media/order/fss.pdf

http://www.faa.gov/air_traffic/publications/ATpubs/AIM/aim.pdf

Air Traffic Organization (ATO)

FAA ATO maintains a nationwide communications net, coordinated with international aeronautical communications services for the control, coordination, and assistance of civil and military air traffic.

Primary FAA resources that may be used for SAR are noted below.

Flight-Following and Alerting Services

ARTCC and ATCT/TRACON facilities provide flight-following service for aircraft on flight plans under instrument flight rules (IFR) or when aircraft are receiving visual flight rules (VFR) advisory service. When ATCT/TRACON facilities provide service and the aircraft becomes overdue they will forward the missing aircraft report to the nearest ARTCC who will alert the RCC. ARTCC facilities providing service to an aircraft that becomes overdue will report the occurrence directly to the appropriate RCC.

FSSs and Flight Service Hubs (FSHs) that provide flight-following service for aircraft on flight plans under visual flight rules (VFR) or who receive a report of a missing aircraft will alert RCC's when these aircraft become overdue.

Radar nets

Radar nets are independent and joint radar sites operated by the FAA and the Air Force. They provide almost complete coverage of the continental U.S., Alaska, Hawaii, Panama, and Puerto Rico. All sites are equipped with IFF interrogators for Air Traffic Control (ATC) use. Most major civil and military aerodromes have short-range terminal radar that may obtain radar contact with distressed aircraft in the vicinity.

National Track Analysis Program (NTAP)/National Track Analysis (NTA)

Radar information in the form of Continuous Data Recordings (CDR) from TRACON facilities or National Track and Analysis Program (NTAP)/National Track Analysis (NTA) from most ARTCC facilities can be requested by RCC's. This information can greatly assist an aircraft search by providing the route of flight and last known position to the extent radar information exists.

NTAP/NTA can retrieve computer-stored radar data up to 15 days old to pinpoint a missing aircraft's last known position. The key item for a successful radar solution is the object's last fix. Many aircraft have been found near the end of their radar track. However, due to the possibility of errors in producing radar tracks, or when multiple tracks exist, search planners should not disregard other search leads.

Distress may not be the only reason an aircraft is lost from radar coverage. Aircraft radar tracks can disappear and reappear more than once along a route of flight due to terrain interference.

NTAP/NTA and other radar information should be requested through the RCC that is coordinating with the radar facility for the release of data.

The more information provided to FAA assets, the more likely the aircraft will be located. To properly process a radar request, the following information should be provided:

- Correlating time and location;
- Direction of travel; and
- Intended destination.

National SAR School

The National SAR School, located in Yorktown, Virginia, is sponsored by the USCG and the USAF. The school offers courses in Land Search Planning and Maritime Search Planning, with instructors from the USCG and USAF. Land Search Planning courses are offered to personnel from government agencies at any level, law enforcement, emergency management, military, CAP, and members of volunteer organizations. There is no tuition charge for U.S. citizens directly involved in SAR.

Further information concerning the Inland SAR Planning course can be found on the National SAR School website: http://www.uscg.mil/tcyorktown/Ops/SAR/Inland/default.asp

Nationally Organized Resources

National Association for Search and Rescue (NASAR)

NASAR focuses on education and certification in introductory and basic SAR skills, search management, search theory, search planning and management of search incidents. The focus of its education is in SAR skills for the field responder and search management. Membership is open to local SAR organizations, local authorities, training institutions and individual SAR responders, both paid and unpaid professionals.

Further information on NASAR can be found at www.nasar.org.

Mountain Rescue Association (MRA)

The MRA is an organization with over 90 government-authorized units in the U.S. and Canada. Members include those who have expertise in technical mountain SAR.

The MRA also specializes in rescue and mountain safety education.

Further information on the MRA can be found at www.mra.org.

National Cave Rescue Commission (NCRC)

The National Cave Rescue Commission (NCRC) received its charter from the National Speleological Society in 1979, and serves as the Society's representative on issues of cave rescue training and operations. It is a volunteer group developed primarily to train and track cave rescue resources throughout the U.S. The NCRC is specifically NOT a functional cave rescue team, rather it provides training and development opportunities for persons and organizations engaged in cave rescue activities. Thus, while many of the persons associated with the NCRC perform rescues, they do this as members of their local rescue squads, civil defense units, or cave rescue groups. The NCRC is a component of the Department of the Administrative Vice-President of the National Speleological Society.

Further information concerning NCRC can be found at http://www.caves.org/ncrc/national/.

National Ski Patrol

The National Ski Patrol is a member-driven professional organization of registered ski patrols, patrollers and others, both paid and volunteer. The NSP supports its members through credentialed education and training in leadership, outdoor emergency care, safety and transportation services, which enables members to serve the community in the safe enjoyment of outdoor recreation.

Further information on the National Ski Patrol can be found at http://www.nsp.org/1/nsp/home.

National Voluntary Organizations Active in Disaster (NVOAD)

National Voluntary Organizations Active in Disaster (NVOAD) is a forum where organizations share knowledge and resources throughout the disaster cycle (preparation, response, and recovery) to help disaster survivors and their communities.

Further information on NVOAD is available at http://www.nvoad.org/.

U.S. SARSAT Program (Overview)

The U.S. SARSAT (Search and Rescue Satellite Aided Tracking) program is an integral part of the International Cospas-Sarsat Programme ("Cospas" is a Russian acronym that stands for "Cosmicheskaya Systyema Poiska Aariynyich Sudov" which translates loosely into "Space System for the Search of Vessels in Distress."). The basic system concept involves the use of distress beacons, satellites, and ground equipment to relay distress location and identification information (referred to as distress alerts) to SAR authorities. SAR instruments are flown on low-Earth polar orbiting (LEO) and geostationary-orbiting (GEO) satellites provided by the United States, Russia, India and the European Organization for the Exploitation of Meteorological Satellites (EUMETSAT). Canada and France provide the SAR instruments for the U.S. LEO satellites. These instruments are capable of detecting signals on the Earth's surface transmitted from distress beacons.

Although the International Cospas-Sarsat Programme ceased satellite processing of 121.5 MHz signals as of 01 February 2009, ELTs still operate on 121.5 MHz or 406 MHz frequencies (by comparison, EPIRBs and PLBs both transmit on 406 MHz only, as well as transmit a low power 121.5 MHz signal for direction finding when activated). 121.5 MHz ELTs only transmit an analog signal that does not contain any information about the distress beacon or user. Alternatively, the 406 MHz distress beacons transmit a digital code that contains information about the type of beacon and possibly the beacon's location (derived from Doppler calculation, Global Positioning System (GPS), or other navigational system). Each 406 MHz distress beacon in the world has a unique identifier. The unique identifier allows for the registration information to be linked to each distress beacon.

After the satellite receives signals from an ELT, EPIRB or PLB, it relays the signals to earth stations referred to as Local User Terminals (LUTs).

The LUT, after computing the location of the distress beacon, transmits an alert message to its respective Mission Control Center (MCC) via a data communication network. The MCC performs matching and merging of alert messages with other received messages, geographically sorts the data, and transmits a distress message to an appropriate SAR authority (RCC or foreign SAR Point of Contact (SPOC)). The distress message may also be sent to another MCC.

Figure 2-1-1 provides an overview of the Cospas-Sarsat System.

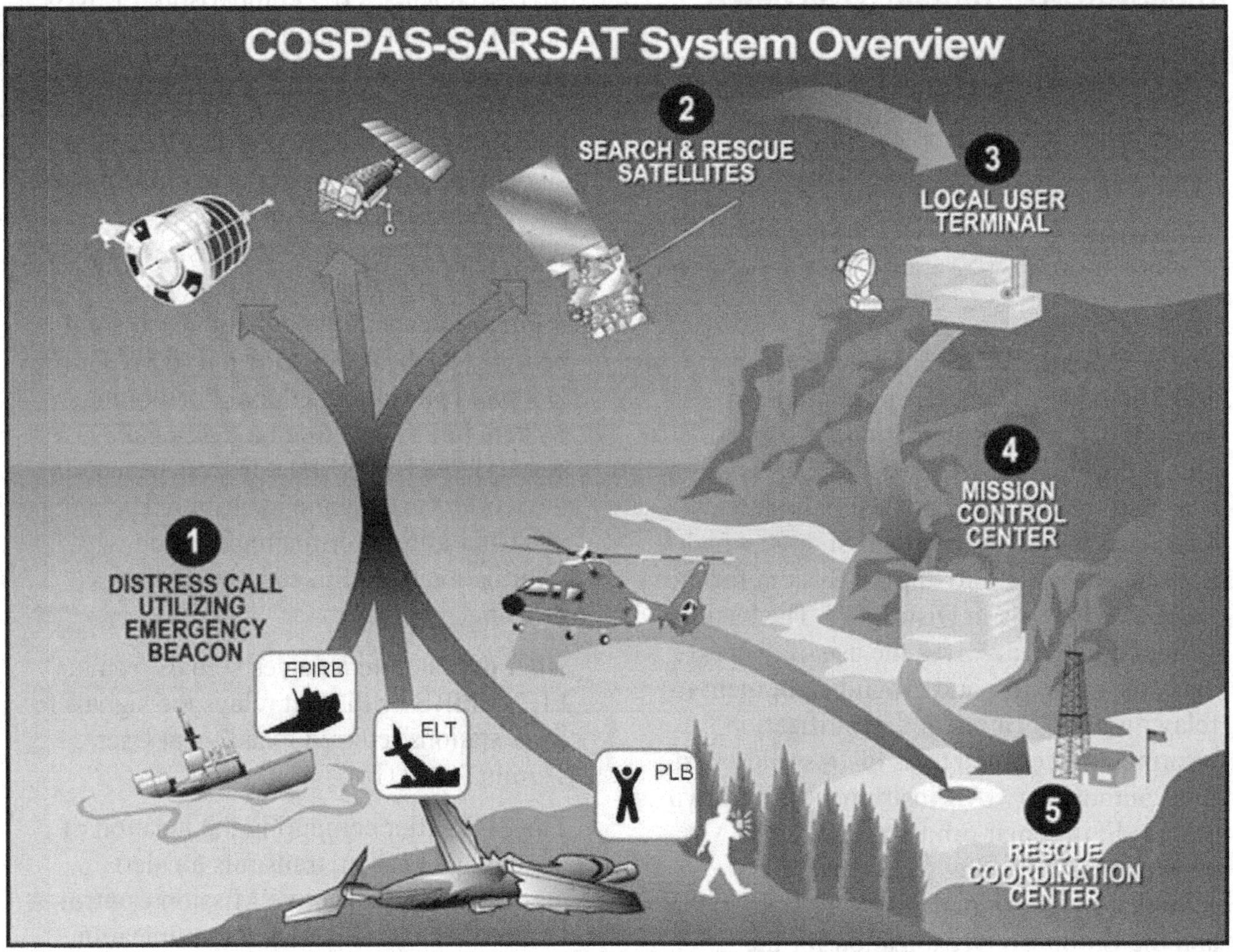

Figure 2-1-1: Overview – International Cospas-Sarsat Program

Cospas-Sarsat System Satellites

The U.S. operates two types of satellites and LUTs to relay distress beacon signals. SAR instruments are carried on board NOAA Polar Orbiting Environmental Satellites (POES) and Geostationary Operational Environmental Satellites (GOES).

POES satellites orbit the Earth at an altitude of approximately 850 kilometers and orbit the Earth once every 102 minutes. The relative motion between the satellites and a distress beacon on the surface of the Earth allows ground processors to use the Doppler Effect to determine the distress beacon's location.

NOAA Polar Orbiting Environmental Satellites (POES)

The GOES satellites orbit the Earth in a geosynchronous orbit along the equatorial plane of the Earth at a speed matching the Earth's rotation. This allows the GOES satellites to "hover" continuously over one position on the surface. The geosynchronous plane is approximately 36,000 kilometers above the Earth, high enough to allow the satellites to view approximately 1/3 of the Earth. Because the GOES (or GEO) satellites stay above a fixed position on the earth's surface, they provide a constant vigil for distress beacon activated within their footprint.

NOAA Geostationary Operational Environmental Satellites (GOES)

LEOLUTs and GEOLUTs

NOAA manages and operates LEOLUTs to track, receive, and process alerts from the POES, European Meteorological Operational (METOP), and Russian Nadezhda satellites. Dual LEOLUTs are located at:

- Andersen Air Force Base, Guam;
- USCG Communication Station, Hawaii;
- NOAA Command and Data Acquisition Station, Alaska;
- Vandenberg Air Force Base, California; and
- USCG Communication Station, Florida.

Dual LEOLUTs at each site allow NOAA to resolve satellite tracking conflicts and provides redundancy in case of failure.

NOAA also manages and operates two GEOLUTs in Suitland, Maryland, to track, receive, and process distress alerts through the GOES East and GOES West satellites. Both sets of LUTs perform error detection and correction on 406 MHz distress beacon messages and automatically generate alert messages to the U.S. MCC (USMCC). Additionally, one LEOLUT and GEOLUT in Maryland serve as backup equipment and test beds.

USMCC

The USMCC (NOAA Satellite Operations Facility, Suitland, Maryland) receives alert data from other international LUTs and MCCs. It matches distress beacon signals to identify those coming from the same source and merges the data to improve position accuracy.

The USMCC appends 406 MHz distress beacon registration information for beacons registered in the U.S., then geographically sorts the data to determine the appropriate recipient (national RCC (Air Force or USCG), foreign SPOC, or other international MCC).

The USMCC filters redundant data and performs system support and monitoring functions. System support functions include relaying SAR instrument telemetry from the Environmental Processing Satellite Center (ESPC) to the Canadian MCC (CMCC) and the French MCC (FMCC) and relaying SAR instrument commands from the CMCC and FMCC to the Satellite Operations Control Center (SOCC).

The U.S. LEOLUTs and USMCC have the capability to detect and locate interference in the 406 MHz band. This information is automatically forwarded to the FCC for further investigation.

This page intentionally left blank.

Section 2-2: SAR Personnel Considerations

SAR Training, Qualification, and Certification

Exercises

Vulnerability Assessments

Critical Incident Stress Management

SAR Training, Qualification, and Certification

A good SAR training program is vital to meeting the SAR system objectives by developing personnel capable of effectively performing SAR operations. Training is critical to performance and safety. Saving lives and reducing the risks to SAR personnel and facilities should be the goal of the training program. Training personnel in making sound risk assessments will help to ensure that these trained SAR professionals and valuable facilities remain safe and available for future SAR operations.

SAR qualification and certification processes are used to ensure that sufficient experience, maturity, and judgment is gained. During a qualification process, the individual must, by demonstration of abilities, show mental and physical competence to perform as part of a SAR team.

Qualification is used to validate an individual's ability to perform certain duties. It is the minimum level of knowledge and skills required to show proficiency in a given task. The qualification program should cover fundamental knowledge necessary for the duties of that position and testing of individuals on the systems they will be required to operate or maintain.

Certification is to authorize an individual to serve in a stated capacity. Certification should be in writing and is the official recognition by the organization that it trusts the individual to use those abilities.

Each SAR entity or authority should establish training, qualification, and certification standards based on the unique environment and challenges of a particular region. These standards should include:

- SAR procedures, techniques, and equipment;
- Assisting in or observing actual operations;
- Exercises in which personnel are trained to coordinate individual techniques and procedures in a simulated operation; and
- First aid instruction that includes demonstrations and exercises led by qualified emergency medical personnel.

Exercises

To reach a high degree of proficiency, SAR authorities should periodically take part in coordinated exercises. Exercises:

- Are invaluable when conducted between multiple agencies that would not normally work together;
- Test and improve operational plans and communications;

- Provide learning experience; and
- Improve liaison and coordination skills.

Generally, there are three levels of SAR exercises:

- *Simple communications exercises* require the least amount of planning and consist of testing communications to ensure its capability for actual emergencies between potential users.
- *Coordination exercises* involve a simulated response to a crisis based on a series of scenarios. SAR managers, planners, and coordination teams are involved, but no fielding of SAR teams. This requires considerable planning and normally one to three days to execute.
- *Full-scale or field exercises* are conducted to deploy and exercise actual SAR resources and agencies. These exercises are the most complex, time consuming, and most expensive, but are extremely effective in evaluating and training involved personnel, SAR organizations, and agencies.

The evaluation process of a SAR exercise is crucial. Those observing and evaluating the exercise must have the expertise in the area they are evaluating and clearly understand what is being evaluated. Identification of weaknesses and development of recommendations for improvement is the outcome.

Vulnerability Assessments

Vulnerability assessments are conducted to accurately assess the local SAR response capabilities within a certain geographic region. It usually contains the following information:

- Area demographics;
- History of the area in relation to past SAR operations;
- Geographic factors; and
- Availability, training, and experience of SAR resources.

The vulnerability assessment should contain a synopsis of what has occurred in the jurisdiction and what may occur under different conditions. A properly prepared assessment will help in future SAR operations by determining type and number of SAR resources available and identifying hazards to SAR personnel. Additionally, this tool can be used to determine the risk to SAR responders.

Critical Incident Stress Management (CISM)

Tragedies, deaths, serious injuries, plane crashes, shootings - these all can seriously affect people who respond in emergencies and encounter highly stressful events. Sometimes an event is so traumatic or overwhelming that people may experience significant stress reactions. These events are known as critical incidents. Critical incident stress is the body's normal reaction to a very abnormal event.

SAR responders can undergo these normal, although uncomfortable stress reactions.

CISM is a comprehensive, integrated, systematic form of stress intervention that is applied to manage critical incident stress after traumatic events. Elements of CISM must be in place before the traumatic event occurs. Pre-incident stress education is essential for organization and individuals. It is also important that standard operating procedures be developed for the organization and individual be developed. CISM teams need to be properly trained.

(Note: Further information on CISM can be found in the CISAR Addendum.)

Section 2-3: SAR Volunteers

Role of Volunteers

Good Samaritans

Unaffiliated Volunteers

Role of Volunteers

Volunteers are usually on the front line helping to rescue disaster survivors, as well as meeting their immediate and long term needs. Some volunteers bring specialized skills and expertise; others offer an extra pair of hands to assist with the work. Volunteers want to help and make a difference.

SAR volunteers may be experienced SAR veterans who are members of federally supported volunteer organizations. Such organizations include FEMA's community emergency response teams (CERT), CAP, or the CGAUX.

In many instances, volunteers who have saved countless lives in some of the worst disasters, are the lone Good Samaritan, who see someone in distress and lends a hand. Countless thousands of victims are saved in big disasters and other, daily difficult situations by people who seize the moment and make a difference.

Volunteers often increase the level of available search effort, which tends to increase the probability of a successful search, or at least potentially reduce the critical time to rescue.

SAR professionals need to learn to use volunteers wisely to help in lifesaving operations. Federal, State, Tribal, Territorial and local SAR authorities need to find ways to organize and train SAR volunteers to be an effective lifesaving force. Be familiar with the capabilities of local volunteer organizations and take steps to facilitate their use.

There are numerous volunteer organizations in the U.S. Persons interested in helping in emergencies should be encouraged to affiliate with one of these volunteer organizations.

Across the U.S., volunteers sacrifice their time and money to help locate lost children, Alzheimer patients, boaters, hunters – everyone and anyone who is in distress and needs assistance. In many instances, SAR volunteers bring much needed assets such as vessels, aircraft, and other critical SAR equipment to a mission that can make the difference in rescuing someone in distress.

Good Samaritans

Eagle, AK, July 29, 2009 - Volunteer helps to prepare the foundation to a new home in Eagle, AK, for a disaster survivor. (Photo: Ben Brennan/FEMA)

A Good Samaritan in legal terms refers to

someone who renders aid in an emergency to an injured person on a voluntary basis.

SAR authorities should be familiar with related laws in the States where they operate or with legal precedence when Federal Agencies provide SAR services.

Usually, if a volunteer assists an injured or ill person who is a stranger, the person giving the aid owes the stranger a duty of being reasonably careful. In some States it is considered an act of negligence if a person does not at least call for help. Generally, a Good Samaritan can help an unconscious victim on the grounds of implied consent. However, if the victim can respond, a person should ask their permission to assist.

Even though some States offer immunity to Good Samaritans, a claim could result if the injuries or illness were made worse by the volunteer's negligence. Statutes typically do not exempt a Good Samaritan who acts in a willful or reckless manner in providing the care, advice, or assistance.

SAR personnel should be aware of certain legal terms that relate to rendering assistance. Lawsuits tend to arise from perceptions of unmet expectations. The *Volunteer Protection Act of 1997* (Public Law 105-19) protects volunteers of a non-profit organization or governmental entity who are acting prudently within the scope of the volunteer's duties and meet certain other requirements. The Act is subject to a number of unprotected exceptions and does not limit the liability of the volunteer organization itself.

Unaffiliated Volunteers

Volunteers unaffiliated with any volunteer organization offer to help or self-deploy to assist in emergency situations and may arrive on scene without fully coordinating their activities. Although unaffiliated volunteers can provide significant resources, because they do not have pre-established relationships with emergency response organizations, verifying their training or credentials and matching them with the appropriate service areas can be difficult. In addition, Federal Agencies cannot accept voluntary services except where authorized by law. Therefore, Agencies should confirm the availability of such legal authorization in advance or encourage individuals to associate with a volunteer organization already participating with the SAR effort.

Houston, TX, September 2, 2005 - Families displaced by hurricane Katrina in New Orleans look for needed items at the Houston Astrodome. Hundreds of unaffiliated volunteers brought clothing, food and other goods to the Astrodome and left them in the parking lot. (Photo: Ed Edahl/FEMA)

The NRF has a Volunteers and Donations Support Annex (http://www.fema.gov/pdf/emergency/nrf/nrf-support-vol.pdf) that describes coordination processes used to support the State during disaster response operations.

State, Tribal, Territorial, and local governments have primary responsibility to develop and implement plans to manage volunteer services and donated goods.

Full use of existing volunteer and donations management resources at the local level is encouraged before seeking assistance of the State or Federal Governments.

Part 3: Communications and Public Relations

This page intentionally left blank.

Section 3-1: Communications

Communications Plans

Interoperability Field Operations Guide (NIFOG)

Communications Support

Distress Beacons: International Cospas-Sarsat Program

Wireless Communications

Satellite Emergency Notification Devices

SAR Action Plans and Reporting

Communications Plans

Communication within and among diverse organizations is essential to an effective SAR response. Communications include all written, spoken, and electronic interaction among all audiences based upon their task-related needs.

No SAR operation can succeed without effective arrangements for communications. SAR communications requirements cannot be met without the right types and amounts of communications capabilities available at the right places, at the right time, installed and supported, with appropriately trained communications personnel. This will not happen without a plan.

A communications plan is a written document that describes who will need interoperable communications capabilities, what will be done with available communications capabilities, how the objectives will be accomplished, and how the success of the communications plan will be measured.

The four important factors in developing the plan are interoperability, priority, reliability, and availability.

- *Interoperability* requires that different SAR entities will be able to communicate with one another during the incident.
- *Priority* deals with the process of handling messages and other communication signals related to SAR.
- *Reliability* is a measure of whether equipment and systems used by persons in distress and providers of SAR services are in good working condition.
- *Availability* refers to the SAR providers' access to equipment.

A communications plan should include the following:

- Radio communications (e.g., terrestrial and satellite, digital and voice frequencies, etc.);
- Print publications;
- Online (and collaborative) communications tools;

- Media and public relations materials; and
- Signs.

National Interoperability Field Operations Guide (NIFOG)

SAR responders require a communications plan that effectively addresses interoperable communications for events of any potential scope. Of course, these plans must be supported with arrangements for the communications capabilities prescribed in the plans.

In assessing communications needs, a good starting place is the National Interoperability Field Operations Guide (NIFOG), which provides a framework for interoperable communications.

NIFOG is a pocket-sized guide of technical reference material for technicians responsible for radios that will be used in SAR response applications. The NIFOG covers regulations on interoperability, available channels, and commonly used emergency frequencies.

The NIFOG is not a replacement for a communications plan, but it does provide the basis for constructing a mission specific plan.

Table 3-1-1: NIFOG Communications Frequencies

Type of SAR	*Frequencies Available*
Land SAR	*Typical Frequencies: 155.160, 155.175, 155.205, 155.220, 155.235, 155.265, 155.280 or 155.295 MHz. If Continuous Tone-Controlled Squelch Systems (CTCSS) is required, try 127.3 Hz (3A).*
Water SAR	*156.300 MHz (VHF Marine ch. 06) Safety and SAR; 156.450 (VHF Marine ch. 09) Non-commercial supplementary calling; 156.800 (VHF Marine ch. 16) Distress and calling; 156.850 (VHF Marine ch. 17) State control; 157.100 (VHF Marine ch. 22A) Coast Guard liaison.*
Coast Guard Auxiliary	*138.475, 142.825, 143.475, 149.200, 150.700 MHz (NB only).*
Aeronautical SAR Coast Guard/DOD Joint SAR	*3023, 5680, 8364 kHz (lifeboat/survival craft); 4125 kHz (distress/safety with ships and coast stations); 121.5 MHz emergency and distress; 122.9 MHz SAR secondary and training; and 123.1 MHz SAR primary. 243.0 MHz AM initial contact; 282.8 MHz AM working.*
Military SAR	*40.50 wideband FM U.S. Army/USN SAR; 138.450 AM, 138.750 AM; 282.8. MHz*
VHF Marine Channels	*6, 9, 15, 16, 21A, 23A, 81A, 83A*

Communications Support

Additional and specialized communication support is often available through other organizations, which may include:

- Radio Emergency Associated Communications Team (REACT);
- Radio Amateur Civil Emergency Services Units (RACES); and
- Amateur Radio Emergency Service.

The SAR Pre-plan can identify communications support agencies in the area of operation.

Distress Beacons: International Cospas-Sarsat Program

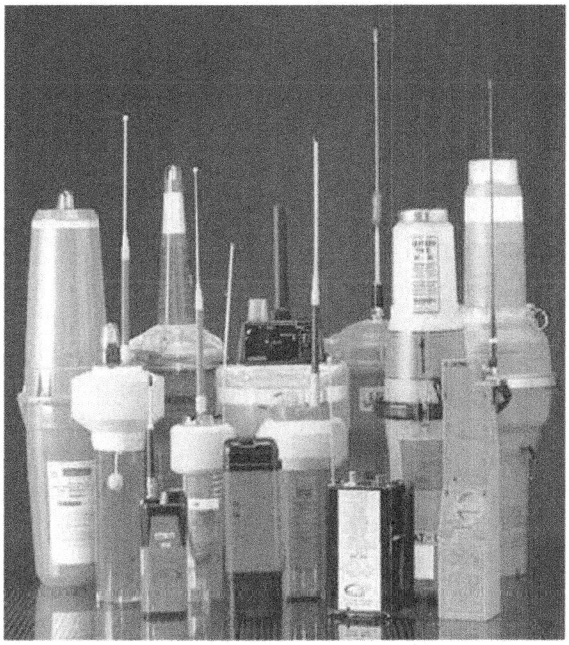

Distress Beacons (Photo: NOAA)

(Note: The NSS includes extensive information about distress beacons and the Cospas-Sarsat system.)

Distress beacons are manufactured to stringent national and international standards and are designed to operate anywhere in the world. When activated, distress beacons notify SAR authorities of a distress situation and the location via satellite.

The beacons transmit distress alerts through the International Cospas-Sarsat Program satellite system on 406 MHz to the appropriate RCC or other SAR point of contact worldwide.

(Note: Older generation beacons transmit on 121.5 MHz. These distress beacons can only be detected by aircraft or other receiver that detects the 121.5 MHz distress beacon signal. Even though these signals are not detected by the Cospas-Sarsat satellite system, they should still be reported to the appropriate aviation or SAR authority.)

406 MHz distress beacons transmit a continuous, low-power 121.5 MHz homing signal. In addition to using the 121.5 MHz signal for homing, many SAR authorities are equipping aircraft and ground SAR responders with the capability to DF on the 406 MHz signal, which transmits for about a half-second every 50 seconds.

406 MHz distress beacons transmit a unique code that allows the alert to be associated with registration data for a particular beacon. Both international and national guidance requires that distress beacons be registered. In the U.S., the FCC requires (by rule) that 406 MHz distress beacons with U.S. codes be registered with NOAA. Distress beacons may be registered in various national or regional databases around the world, or in the Cospas-Sarsat International Beacon Registration Database (IBRD). When a 406 MHz distress beacon is activated, the registration data for the U.S. coded beacon is forwarded to the appropriate RCC along with distress alert.

RCCs receive many distress beacon alerts every day, and should be familiar with the format and contents of the alerts and where to retrieve associated registration data. The most important registration data is the identity of the distress beacon owner and associated owner's emergency contact information.

More information on distress beacon is available at: www.sarsat.noaa.gov and www.Cospas-Sarsat.org.

The types of distress beacons include the following:

- *Emergency locator transmitters (ELTs)* – emergency distress beacons that most U.S. civil aircraft are required to carry. These distress beacons alert via satellites on 406 MHz or via audibly to overflying aircraft or ground stations monitoring 121.5 MHz. ELTs are designed to activate on the impact of an aircraft crash to indicate the crash location, as well as manually;

- *Emergency position-indicating radio beacons (EPIRBs)* – emergency distress beacons that ships, boats and other marine craft carry that alert via satellites on 406 MHz. EPIRBs are not intended to be used on land, can be activated manually, or upon submersion in water;
- *Personal locator beacons (PLBs)* – multi-environment emergency distress beacons individuals carry that alert via satellites on 406 MHz.

Some distress beacons (e.g., location-protocol distress beacons) can transmit a GPS position, either from an internal GPS chip installed in the distress beacon, or from navigation equipment aboard an aircraft or vessel. GPS positions are very accurate, but the accuracy is limited by the number of bits available in the distress beacon to transmit the position.

Regardless of whether the distress beacon has the ability to transmit its position, the position will be calculated and forwarded with the alert by the satellite ground system using a Doppler technique, and is usually accurate to within 3-5 km.

Note: See Appendix B for more information on how Cospas-Sarsat satellites locate beacons.

Wireless Communications

Wireless communication (e.g., cellular, other mobile systems, etc.) usage is growing rapidly, and the public is increasingly relying on cell phones in conjunction with, or sometimes instead of, more suitable or dedicated means of calling for assistance. However, in some cases, cell phones can be a reliable means of notifying the SAR system of a distress situation.

Due to the limitations of cell phones, SAR checklists should include the following additional items for incidents involving cell phone use:

- Caller's complete cellular telephone number including area code;
- Type of cell phone;
- User's cellular service provider or carrier;
- Whether or not other means of communications are available;
- Remaining battery strength in cell phone;
- Establish a communications schedule or require the caller to call back at a scheduled time if possible;
- Ensure user understands that the cell phone, if there is sufficient battery strength, must not be turned off so that further communications can be established;
- Ask if the user has an alternate power source available;
- Ask if the user has the ability to send and receive text messages to conserve battery life; and
- Obtain a good alternative point of contact.

Most cellular service providers offer some of the following services to assist in locating the origin of cellular calls:

- *Call Trace/Cellular Tower Locator* – As long as there is a connection, the carrier's technician can determine which cell tower is receiving the call, and if power and antenna height are available, an approximate arc of distance from the cell tower.

- *Call Trace Modified* – After the call is initiated and the technician is notified, the caller can be instructed to call back at a specified time and the technician can determine through the use of signal strength at several cell sites, a more accurate probable position of the caller.
- *Cell Traffic Recording* – A carrier can determine the cell location of the last call placed by the subscriber given the cellular telephone number.
- *Tap* – This function provides notification when a call is made from the user's phone; beneficial in overdue cases.

Satellite Emergency Notification Devices

The emergence of several non Cospas-Sarsat type certified Satellite Emergency Notification Devices (SENDs) warrants awareness by SAR authorities. These devices are readily available to the general public and require an annual subscription to enable all the features. Their "911" function does not alert traditional Federal SAR services. These alerts are forwarded to a commercial call center which uses a Public Safety Answering Point (PSAP) database to alert police, firefighting, and ambulance services in the alert area. These agencies may or may not be the agency having jurisdictional authority for SAR in that area.

SAR Action Plans and Reporting

The SMC/IC should develop a search action plans (SAP) and a rescue action plans (RAP) as appropriate. These plans are provided to the OSC and SAR facilities on scene and should include:

- A summary of the on scene situation, including the nature of the emergency;
- What to search for (search object(s));
- Weather information;
- Summary of SAR facilities on scene;
- A listing of the appropriately-described search area(s) and sub-areas that can be searched by the SAR facilities in the allotted time;
- On land or flooded areas, georeferencing grids may be added to designate positions; and
- The assignment of communications frequencies to include primary and secondary control channels, on scene, monitor and press channels, and special radio procedures, schedules, or relevant communication factors.

Reporting requirements differ widely by unit. Each SAR entity should establish the reporting requirements for their units.

This page intentionally left blank.

Section 3-2: Family, Media, and Public Relations

Media and Public Relations

Family Support Services

Media and Public Relations

The public should be informed during SAR operations, within the limits of confidentiality, of SAR system actions. The potential benefits of early release of information include:

- Additional information from the public, leading to more effective use of SAR resources;
- Fewer time-consuming requests from the news media; and
- Reduction in inaccurate public speculation about the SAR mission.

A SAR operation often creates great interest with the general public and with radio, television, and newspapers. Contacts with media are normally the responsibility of the SMC/IC through the Public Information Officer (PIO). It is important that a good relationship between the media and the SAR organization is established to ensure that information reaching the public is factual and complete.

In order to ensure the formulation of a consistent and controlled message to the public, the PIO should be designated as the focal point for the releases of information relating to SAR operations. Press releases or media conferences can be used as an early release of information, a public update on progress, and as a final release summarizing the entire case after SAR operations are concluded. All information released by the PIO should normally be approved by the SMC/IC and appropriate authorities, and contain only factual information.

Once initial public information has been released, the PIO should consider programming and advertising regular and frequent updates in order to address the needs of the media. These could take the form of further press releases or holding press conferences. A press conference gives the PIO the opportunity to initiate the following actions:

- Give information;
- Give interviews;
- Answer questions; and
- Give the media controlled opportunities to obtain video footage, photographs, and audio for broadcast use.

Family Support Services

Families are an important part of the SAR mission. How they are treated can determine not only how they could help the SAR effort, especially in narrowing down the search area, but also how they will represent the SAR agency when the media calls.

Assign a single 24- hour family liaison point of contact. This is usually a staff position under the IC. This person should possess two important qualifications: ability to empathize with the family members and sufficient knowledge of search management to explain strategies and tactics to the family. Additionally, the person assigned should be available for the incident's duration in order to create the rapport and trust needed in establishing lines of communication.

The family liaison must be able to give the family needed emotional support by providing honest reassurances and progress updates. To minimize stress and worry, briefings to the family should be:

- Scheduled to meet the needs of the family and the search debrief cycle;
- Positive, but not inaccurate or leading to false hopes;
- Informative; and
- Professional, instilling confidence that all possible actions are being taken.

In general, the search agency should not ignore the family. Assign a family liaison officer, and keep them constantly informed.

Part 4: Operations

This page intentionally left blank.

Section 4-1: Awareness and Initial Actions

General

Response to a SAR incident normally proceeds through five stages (See Figure 1-6-1):

- Awareness;
- Initial actions;
- Planning;
- Operations; and
- Conclusion.

For a particular incident, response may not require performance of all five stages, or stages may overlap with elements of several stages taking place at once. The first responder on scene or the IC in the early stages of a response must be ready to guide the SAR incident through each stage, taking the correct actions at the right time. The goal is to find the person(s) missing or perform the rescue quickly, efficiently, and safely.

The IC must be prepared to take decisive action in an effort to locate the search object quickly. ICs and first responders should be knowledgeable and trained on initial actions to take (or consider) in order to start a search "on the right foot".

Actions on Arrival at SAR Incident

SAR responders may arrive on scene during the initial response or at a developing search after some initial actions are in progress or have already taken place. SAR responders should have in their pre-plan actions to take when arriving to minimize response time. Items to consider include:

- Upon arrival at a developing search, leaders should check in with the current IC for briefing on actions taken so far, by whom, and what actions are currently being taken;
- Establish the ICS;
- Assign personnel on scene and/or early arriving personnel to cover hazards in the area, such as: approaches to cliff areas, swift water, deep water, major roadways, etc. (especially important in cases involving children or vulnerable adults);
- Complete the Lost Person Questionnaire (Appendix E);
- Complete the Missing Person Checklist (Appendix F);
- Begin investigation and protect the LKP or PLS;

- Establish Incident Command Post;
- Assign a staging area manager and set up staging area;
- Establish check in procedures and check in all SAR personnel on-scene (ICS Form 211);
- Analyze mission and prioritize tasks; and
- Immediately after a SAR incident begins, plan for personnel relief (failure to relieve fatigued personnel could lead to critical errors in search operations and planning).

Implementing ICS

Using ICS principles, designate a Command and General staff. Staff should be based on required functions.

A single person may carry out more than one function, but a single function should not be held by more than one individual unless the additional person is to serve as relief.

At a minimum, the ICS functions that should be considered include:

- Incident Command (IC);
    - Safety Officer;
    - Public Information Officer;
    - Liaison Officer;
- Operations Section;
- Planning Section;
- Logistics Section;
- Communications Unit; and
- Other Liaisons as appropriate.

Under the ICS principle of Unified Command, all agencies having jurisdiction may have position in the Incident Command.

Missing Person Profile

Using the Initial Missing Person Checklist, begin to construct a missing person profile:

- Interview direct family members, friends who were with the subject at the time of the disappearance, extended family members, and individuals who may have knowledge of terrain, map features, unmarked trails, and other landmarks to gain as much information as possible.;
- Compile the information to complete a Lost Person Questionnaire;
- Compile all known intelligence (searching and planning) data from all previously completed tasks;
- Determine search urgency;
- Based on all the information, determine a theoretical, statistical, and subjective distance the subject could have traveled, then select an area to target with initial search actions; and
- Consider calling out additional resources early in the search to increase the areas covered by the initial search actions and improve the chances of finding the subject quickly.

Initial Search Actions

The IC will need to plan and conduct a Rapid (Reflex/Hasty) Search, which may include trackers, canines, and sound teams, as well as tasks such as perimeter patrols and trail checks as indicated by the situation and missing person profile developed.

- For searches where the LKP/PLS is a residence or structure, once personnel have been assigned to cover hazards, perform a thorough check of all buildings in the vicinity of the LKP/PLS to include attics, rafters, lofts, and basements of all structures, as well as tall grass areas, scrub, and wood lines around the perimeter of the property;

- Establish confinement: Perimeter road patrols, if used, should strive for visual checks of roads/ditches in the area covered by Rapid Search at least once every 30 minutes;
- Perform trail checks or trail-running and have searchers scan the trails and environment/terrain to either side of trails;
- Keep one or several people available to follow up on intelligence sources or leads as they become available (Law enforcement personnel may be a good choice for this task);
- Send law enforcement personnel (if available) to check nearby businesses such as taverns or stores for the subject, and to gather additional intelligence, as well as check nearby homes;
- Consider sound searches;
- Tracking and sign cutting are important tasks during initial response and Rapid (reflex/hasty) Search phase. As people walk, they will leave signs of their passage. Skilled trackers can acquire, age, and follow sign. Trackers can work with canine teams without the two teams interfering with each other; and
- Canine resources may provide significant clues for the search.

During the Rapid Phase;

- Brief all field teams;
- Maintain confinement;
- Perform Rapid Search actions as called for by the mission profile;
- Debrief all returning teams;
- Record all efforts (for example, this may be accomplished using ICS 214); and
- Analyze mission profile to prioritize Rapid (Reflex/Hasty) Search tasks.

After all Rapid (Reflex/Hasty) Search resources have deployed to the field, the focus of activity in the command post should shift to the Communications and Situation functions, who will handle radio traffic, position and status reports from the field. While the Rapid Search is taking place, the General Staff should begin planning for the next operational period.

Lost Person Behavior

Studies on how missing persons act and where they are located can be useful to SAR planners and ICs. The likely behavior of the missing person is an important consideration in establishing search efforts.

While lost person behavior studies provide numbers and percentages, their key use is in the information they provide about the varieties, similarities, differences, possibilities, and impact of how different people behave when lost.

Data may help make decisions regarding determination of a distance beyond which there is no reason to search, in the absence of more clues. Some of the information may help determine what types of resources to use and tactics, as well as likely routes of travel and likely locations to search early and quickly. Distance information assists with confinement efforts.

Lost person behavior studies by William Syrotuck, Ken Hill, Barry Mitchell, and Robert Koester are all readily available. The largest is the International Search and Rescue Incident Database (ISRID) developed by Koester. The data refines the statistics for missing persons, and is divided into temperate and dry climates, mountainous, non-mountainous, and urban regions. The ISRID includes statistics on dispersion angles and find locations.

Based on these studies, lost people tended to exhibit characteristics that could be categorized according to their age and the activity they were engaged in. Different "categories" of people tended to travel different distances from the places where they were last seen. Where people were found (distance from Initial Planning Point - IPP) was not necessarily related to how the people arrived there or how long the people traveled.

When selecting a category, refer to subject profiles in the studies. Consider the mental state, purpose, intention, and motivation of the missing person, as well as their physical capabilities (e.g., Does the person know he/she is lost?).

The ISRID has sufficient data to report on 41 subject categories plus overdue aircraft, and are listed in Table 4-1-1 below.

(Note: An example ISRID Lost Person Behavior Category - Dementia (Alzheimer's), can be found in Appendix G.)

Table 4-1-1: International Search and Rescue Incident Database Categories		
Abduction	Despondent	Snowboarder
Aircraft	Gatherer	Snowshoer
Angler	Hiker	Substance Abuse
ATV	Horseback Rider	Urban Entrapment
Autistic	Hunter	Vehicle (Missing Vehicle)
Camper	Mental Illness	Vehicle (Four-Wheeled Drive)
Caver	Mental Retardation (Intellectual disability)	Vehicle (Abandoned Vehicle)
Child 1-3	Mountain Biker	Water (Powered Boat)
Child 4-6	Other (Base Jumper)	Water (Non-Powered Boat)
Child 7-9	Other (Extreme Sports)	Water (Person In Water – Flat Water)
Child 10-12	Other (Motorcycle)	Water (PIW – Moving water)
Child/Youth 13-15	Runner	Water (PIW – Flood Stage)
Climber	Skier – Alpine	Worker
Dementia (Alzheimer's)	Skier - Nordic	

Lost person behavior statistics are a useful tool for initial and extended search efforts. Statistics that give probabilities of where the subject might be located and an overview of types of behaviors, likely actions, and intentions of the lost person, can help determine areas where planners may wish to focus search efforts and assist in determining a desired search area.

Section 4-2: Land SAR Missions (Non-Aircraft)

Types of Operations

Use of Federal Resources

SRU Facility Reporting

Types of Operations

Typical land SAR operations (other than aeronautical SAR) include:

- Missing person search;
- Rescue;
- Transport of SRUs;
- Mercy;
- Medevac; and
- Mass rescue operations (MROs).

Normally, State and local SAR authorities have the primary responsibility for non-aircraft land SAR missions. This can be affected by agreements, and is not intended to ever preclude prompt and effective Federal action to save lives when necessary.

Patient and medical transports, while not actually SAR operations, may be supported and coordinated by Federal RCCs. There is no obligation to provide such services, and they should not be provided in a manner that competes with private or commercial interests.

Use of Federal Resources

There are many constraints regarding the use of Federal SAR resources, particularly for land SAR. The following are some factors that will be considered in evaluating requests for Federal SAR assistance:

- State, Tribal, Territorial, and local SAR authorities usually provide SAR resources for land SAR operations. These authorities should normally request Federal SAR resources only when non-Federal resources are not available, or are overwhelmed (National Guard resources are among those that can be tasked by a State);
- Urgency - Federal SAR resources should not be used for convenience. A request to use Federal SAR resources will only be considered when there is an actual person(s) in distress, an immediate threat to life, limb, or eyesight, or likely undue human suffering (An exception may be when a SAR team would be put at risk to accomplish a mission without the aid of Federal resources.);
- Nature of SAR event, the presence and/or characteristics of the defined search area, and the type of subject may require the use of Federal SAR resources;
- There is an acceptable level of risk to the Federal SAR resource;
- Consistency with legal restrictions (e.g., Posse Comitatus Act, Stafford Act, and Economy Act);
- Federal SAR Resource availability - Federal SAR resources are evaluated for distance from the scene, special equipment requirements, terrain factors, urgency of the situation, and conflicts with military or other priority mission requirements; and

- Requests for Federal SAR resources should normally be for a capability, not for a specific type of asset.

Appendix H provides additional information for evaluating Federal SAR resource requests.

SAR Facility Reporting

During all missions, SAR facilities should submit a report to the SMC at specified intervals. This interval may vary depending on the tempo, size or complexity of the SAR operations, but should always be submitted at least daily.

Section 4-3: Missing Aircraft SAR Missions

Overdue or Missing Aircraft

FAA Order JO 7110.10U defines an "overdue" aircraft:

- An aircraft on an IFR flight plan is considered overdue when neither communication nor radar contact can be established and 30 minutes have passed since its expected time of arrival (ETA) over a specified or compulsory reporting point or at a clearance limit;
- An aircraft on a VFR flight plan is considered overdue when it fails to arrive 30 minutes after its ETA and communications or location cannot be established; or
- An aircraft not on a filed flight plan is considered overdue at the actual time a reliable source reports it to be at least one hour late at its destination. Based on this overdue time, apply the same procedures and action times as for aircraft on a flight plan. When such a report is received, verify (if possible) that the aircraft actually departed and that the request is for a missing aircraft rather than a person. Refer missing person reports to the appropriate authorities.

Notification Sources

The following are FAA notification sources:

- *Domestic Event Network (DEN)* - The DEN is an interagency teleconference that may provide timely notification to the appropriate authority if there is an emerging air-related problem or incident within the continental U.S.;
- *Aeronautical Information System-Replacement (AIS-R)* – The AIS-R is a web-based system designed to collect and distribute weather, flight plan data, NOTAM messages, pilot report messages, and other operational information to all ARTCC and FSS facilities. AIS-R serves as a backup system in the event the primary Flight Service Operational System is unavailable;
- *Hazardous Area Reporting Service* – FSS facilities when requested will provide Lake, Island, Mountain, or Swamp reporting services which require a radio contact with the pilot every 10 minutes (or at designated position checkpoints) while crossing a hazardous area. If contact with the pilot is lost for more than 15 minutes, search and rescue authorities will be alerted; and
- *Other FAA sources* - Organizations monitoring distress/beacon and other frequencies (e.g., ARTCC, FSS, ATCT/TRACON, transiting aircraft, etc.) may receive a distress report/distress beacon signal; this notification comes to an RCC via telephone, fax, or email.

RCCs receive all 406 MHz emergency locator transmitter (ELT) or personal locator beacon (PLB) distress alerts from the Cospas-Sarsat satellite system via the USMCC.

The RCC is normally notified of family concerns, local sightings and other reports via local law enforcement or emergency response authorities.

Investigation Searches

(Note: Common information sources for investigation searches are discussed in Section 1-4: Rescue Coordination Centers.)

The FAA is responsible for initiating actions to locate overdue or unreported aircraft by generating a request for information from the Departure Station/Facility (QALQ), Information Request (INREQ), and Alert Notice (ALNOT) messages.

Internally, FAA ARTCCs serve as the central points for collecting information, for coordinating with the RCC, and for conducting a communications search by distributing any ALNOTs concerning:

- Overdue or missing IFR aircraft;
- Aircraft in an emergency situation occurring in their respective area; and
- Aircraft on a combination VFR/IFR or an airfield IFR flight plan and 30 minutes have passed since the pilot requested IFR clearance and neither communication nor radar contact can be established with it (For SAR purposes, these aircraft are treated the same as IFR aircraft.).

Flight Service Stations (FSS) serve as the central points for collecting and disseminating information on an overdue or missing aircraft which is not on an IFR flight plan.

When a VFR aircraft becomes overdue, the FAA will attempt to locate the aircraft by checking the destination airport, all adjacent airports that could accommodate the aircraft, and air traffic control facilities near the destination airport. If this communications search does not locate the aircraft, the FAA will initiate a QALQ (Request for Information from the departure station/facility) message and, when appropriate, to the Direct User Access

Terminal (DUAT) vendor or facility where the flight plan information is on file.

(Note: When an IFR aircraft becomes overdue the FAA will immediately issue an ALNOT.)

If the reply to the QALQ is negative, or the aircraft has not been located within 30 minutes after it becomes overdue, an INREQ message to the departure facility, base of operations (BASOPS), en route FSSs, and ARTCCs along the route will be transmitted. The facilities receiving the INREQ will search for the aircraft by checking facility records and all airports within their jurisdiction along the proposed route of flight that could accommodate the aircraft.

The RCC should compare COSPAS-SARSAT distress beacon messages with INREQ and ALNOT messages and advise the INREQ/ALNOT originator or lead FAA authority of any correlations.

If the replies to the INREQ are negative, or if the aircraft is not located within 1 hour after transmission of the INREQ, whichever occurs first, an Alert Notice (ALNOT) will be transmitted to the FAA Regional Office and those facilities within the search area. The search area is normally that area extending 50 miles on either side of the proposed route of flight from the last reported position to the destination. The search area may be expanded to the maximum range of the aircraft at the request of the RCC or by the ALNOT originator. In addition, DUAT vendors and the RCC will be notified.

Upon receipt of an ALNOT, each FSS whose flight plan area extends into the ALNOT search area will conduct a communications search of those flight plan area airports which fall within the ALNOT search area that could accommodate the aircraft and were not checked during the INREQ search. Each ATCT and ARTCC within the ALNOT search area will review their records in an attempt to locate the aircraft. The results of the search will be forwarded to the ALNOT originator and to the RCC. All possible sources of information concerning the missing aircraft or person should be contacted. This is normally conducted after the Uncertainty Phase but during the Alert Phase.

The RCC normally opens an active "Incident" and assigns an SMC once the INREQ communications search has been completed.

The SMC should:

- Compare distress beacon messages and other notification sources with the INREQ message;
- Review weather over the filed or suspected route of flight;
- Review available aircraft registry/persons on board (POB) data to assess the capabilities of the aircrew and aircraft; and
- Evaluate the impact of the environment to help formulate a rough search area if the aircraft is not found during PRECOM.

Table 4-3-1 on the next page provides a summary of QALQ/INREQ/ALNOT requirements, emergency phases, and time requirements.

Table 4-3-1: Summary of Missing Aircraft Investigative Processes			
	QALQ/INREQ	**ALNOT**	**SAR Mission**
Time	FAA initiates a QALQ or INREQ when a VFR aircraft has not been located within 30 minutes after becoming overdue.	If INREQ is negative, or if the aircraft is not located within 1 hour after transmission of the INREQ, or the aircraft is on an IFR flight plan and becomes overdue (whichever comes first), the FAA will initiate an ALNOT.	If the search fails to locate the aircraft, or if 2 hours has elapsed since the ALNOT was transmitted, the Distress Phase commences.
Emergency Phase	Uncertainty Phase	Alert Phase	Distress Phase
Investigation Search	Preliminary and/or flight plan area search begins.	Extended communication and record search begins.	Investigation fails to confirm non-distress no later than 2 hours after the ALNOT is transmitted.
Area to be investigated	Check all airports and facilities along the filed or suspected route of flight.	Check airports and facilities within 50 miles on either side of the filed or suspected route of flight.	Based on available information, the SMC develops the search area; conducts SAR operation.

Search Mission

If the extended communication and record search fails to locate the aircraft, or if two hours has elapsed since the ALNOT transmission, whichever occurs first, the ALNOT originator should contact the RCC, which will mark the beginning of the "Distress Phase."

The RCC should open an active mission and coordinate SAR operations when:

- A distress situation is confirmed;
- Investigation fails to confirm non-distress no later than two hours after the ALNOT transmission;
- One or more distress beacon notifications correlate to the filed or suspected route of flight; or
- Radio or radar coverage drops off in an area of known radio or radar coverage.

The above are not an inclusive list of criteria. The SMC must use good judgment to modify, combine, or bypass SAR stages and procedures to cope with unique, unusual, or changing circumstances. Nothing in this Addendum is intended to impede the decision to open a SAR mission.

The RCC should:

- Review the lead and supporting SMC responsibilities;
  - o Normally, the RCC is SMC for general aviation (GA) interstate flights, DoD aircraft, commercial aircraft, and missions of national concern;
  - o For all other missions, the SMC is determined by individual Federal-State agreements; and
  - o Supporting SAR agencies may provide SAR capabilities and assets at the request of the SMC;
- Alert the appropriate State SAR authority and other responsible agencies per the Federal-State agreement (During

GA interstate flight missions, supporting SAR authorities will be alerted for each State which the aircraft's filed or suspected flight route transits);

- Coordinate use of any Federal SAR assets such as DoD, CAP, Coast Guard, or ICE facilities;
- Coordinate Federal non-aircraft capabilities, including:
    - Radar Forensics (e.g., CDR/NTAP);
    - Cell phone triangulation;
    - Use of personal information;
    - National Alert Warning System (NAWAS);
    - Weather reporting;
    - Coordination of controlled airspace;
    - NOTAM/TFR coordination;
    - Distress beacon alert correlation; and
    - Information Awareness/Assessment (IAA).

A "lead" is information pertaining to the search mission; leads come to the attention of the RCC in a number of ways, such as from the results of the Federal non-aircraft capabilities listed above, local law enforcement and first responders, and eyewitness accounts. All leads should be recorded and communicated to the appropriate State SAR authority, IC, other investigatory agency, or investigated and resolved by the SMC.

Distress Beacon Operations

The AFRCC and AKRCC are responsible for the management of inland distress beacon searches (121.5 MHz, 243.0 MHz and 406 MHz ELTs for aircraft; 406 MHz EPIRBs for the maritime environment; and 406 MHz PLBs for personal use in multiple environments). RCCs coordinate the use of Federal and DoD assets to search for, locate, and silence any activated distress beacon within the continental U.S. and Alaska.

Beacon Response Policy

When the RCC receives a distress beacon alert and/or audible distress beacon alert (airborne and ground), the investigation should be expanded as necessary to aggressively pursue the cause of each alert and dispatch SAR resources to assist as circumstances require.

Audible airborne reports, typically received from the ARTCC, normally include the following information:

- Aircraft altitude;
- Where and when the signal was first heard;
- Where and when maximum signal strength was heard; and
- Where and when the signal faded or was lost.

The RCC should correlate reports with known open incidents and missions and with any Cospas-Sarsat alerts. If no correlation exists, a new incident is opened and an investigation begins.

Ground reports, typically received from airfields and FSSs, normally include:

- Time heard;
- Frequency; and
- Signal strength.

The RCC should correlate reports with known open incidents and missions and any Cospas-Sarsat alerts. If no correlation exists, a new incident is opened and an investigation begun. Airfield managers may have the ability to locate the signal and silence it. When this capability does not exist, the RCC should immediately open a mission to silence the distress beacon to prevent masking potential signals from other distress beacons in the area.

SAR Process

Incident Stage

During the incident stage, the RCC should determine whether a distress or overdue situation exists.

Only enough credible data to begin an investigation is required to open an incident. An event will remain at the incident stage until it can be validated that a person(s) is actually at risk. Events that typically warrant an incident include:

- First or composite distress beacon alert;
- Request for assistance from another RCC or SMC, e.g. State Troopers, National Park Service, USCG, military police, first responders or a civilian Medevac company; and
- Notification of an overdue aircraft from the FAA.

Mission Stage

An incident is normally upgraded to a mission when:

- A distress situation is confirmed and a SAR facility is tasked;
- A SAR facility needs to be dispatched to determine whether a distress situation exists;
- A medevac appears to require a SAR facility to help save life, limb or eyesight, or to mitigate undue suffering; or
- An ALNOT is issued by the FAA.

During the mission stage, the RCC should actively maintain control of Federal assets, and the RCC should monitor the mission process through completion.

SAR asset selection is based on matching asset capabilities of available military and civilian assets with the mission requirements.

SAR Facility Reporting

During all missions, SRUs should submit a report to the SMC at intervals specified by the SMC. The interval may vary depending on the tempo, environmental conditions, size or complexity of the SAR operations, but should always be submitted at least daily.

Section 4-4: SAR Resource Strategies and Tactics

Identify Resources

Search Patterns

Passive (Indirect)/Active (Direct) Ground Search Team Strategies

Canine SAR Teams

Mounted, All Terrain Vehicles (ATV), Snowmobile, and Bicycle SAR Teams

Tracker SAR Teams

Aeronautical SAR Resources

Boat SAR Teams

Dive SAR Teams

Identify Resources

The SMC/IC must identify the SAR resources available for use in an incident, and coordinate use of the resources to provide the most appropriate and capable response. To do so, the SMC/IC must be familiar with the operational capabilities of water, land, and air based SRUs in their area.

To assemble the most appropriate search resources for a SAR incident, the SMC/IC should:

- Examine the search area environment (e.g., size, terrain, vegetation, weather, time of day, etc.);
- Develop likely scenarios for what might have happened (likely scenarios considered allow search planners to predict the types of clues they expect to find within the search area);
- Based on the search area characteristics, select the search unit(s) best suited to conduct the search;
- Based on the types of clues expected during the search, select the best sensor(s) for detecting those clues; and
- Compare the capabilities of available SAR resources to the needs and conditions of the search and the desired search unit and sensors, then select and assign the most appropriate resources to conduct the search.

Search Patterns

After determining *where* to search, the SMC/IC must determine *how* to search. The process of determining how to search should be based on *objectives* for the incident and each operational period.

Once objectives have been determined, *strategies* to achieve the objectives are selected.

Strategies lead to *tactics* employed to implement the strategy. The tactic is how you plan to have your searchers execute your strategy. For example, Rapid Search tactics could include route or points of interest (spot) searching. Possible tactics for a "confinement" strategy could include road/trail blocks or track traps.

By developing objectives for a search, the SMC/IC can determine the strategies and tactics needed to achieve the objectives.

Passive (Indirect)/Active (Direct) Ground Search Team Strategies

Two major ground search strategies include:

- Passive (Indirect) Approach.
- Active (Direct) Approach.

Passive (Indirect) Approach (Table 4-4-1) includes strategies, tactics, and considerations on gathering information regarding the subject, why the subject is missing, containing/confining the subject, and attracting the subject out of the search area.

Active (Direct) Approach (Table 4-4-2) includes strategies, tactics, and considerations for active searching using a variety of search resources.

After determining tactics, individual search resources can be assigned to accomplish the desired tactics.

What follows are several Tables and Figures that provide strategies, tactics, and considerations for various types of SAR resources used in the conduct of search operations.

<table>
<tr><th colspan="3">Table 4-4-1: Ground SAR Team Strategies, Tactics, and Considerations
Passive (Indirect) Approach</th></tr>
<tr><th>Strategy</th><th>Tactics</th><th>Considerations</th></tr>
<tr><td rowspan="5">Confinement
(Figure 4-4-1)</td><td colspan="2">Keep subject within search area, detect if subject has left search area; locate subject on roads or trails.</td></tr>
<tr><td>Road/trail blocks</td><td>• Established on all roads/trails leading into and out of area; and
• May be used in conjunction with string lines and signs.</td></tr>
<tr><td>String lines/signs</td><td>• If possible, combine with attraction if possible;
• Placed waist high with paper arrows pointing toward camp or road; and
• Ensure in the language of missing subject, age appropriate.</td></tr>
<tr><td>Track traps</td><td>• Used to detect if subject has traveled through the area;
• Brush off bare areas to detect footprints; and
• Stream/shore banks and areas of loose dirt provide natural track traps.</td></tr>
<tr><td>Perimeter search</td><td>Trackers can sign cut boundaries that tend to collect signs (fence lines, road edges, trails, stream beds, etc.).
• Secure the confinement perimeter.
• Limit priority segments from the search area (if the missing subject did not cross some boundary, little need to search there).
• Establish a new LKP and a direction of travel.</td></tr>
<tr><td rowspan="4">Attraction
(Fig 4-4-1)</td><td colspan="2">Encourage the missing person to find the searchers (assumes the missing subject is mobile and able to follow signals to a place of safety).</td></tr>
<tr><td>Lookouts</td><td>• Using Lookout or Observation towers (Fire Towers) to look for subject;
• May use handheld thermal imagers or binoculars;
• A ladder, aerial, or tower truck may be used as a portable lookout;
• Also due to the lookouts height it would attract the subject; and
• Look-over's, scenic views, bridges, are forms of lookouts.</td></tr>
<tr><td>Helicopter/Airplane Flyovers</td><td>• Aircraft flyovers may attract the missing subject to come out to a clearing or open area and be seen; and
• Use of public address systems mounted on aircraft may also attract subject.</td></tr>
<tr><td>Public Address Sound Devices (Sirens)</td><td>• Use of public address systems to call for the subject; and
• Must be stationary so the mobile subject can go toward the sound.</td></tr>
<tr><td rowspan="4">Investigation</td><td colspan="2">Obtaining missing person information and determining why the missing subject went missing.</td></tr>
<tr><td>Missing Person Questionnaire</td><td>• Complete missing person questionnaire.</td></tr>
<tr><td>Lead Follow-up</td><td>• Follow-up on leads/clues to determine relevance to the missing person.</td></tr>
<tr><td>Interviews</td><td>• Interview family, friends, neighbors, doctors, co-workers, etc.</td></tr>
</table>

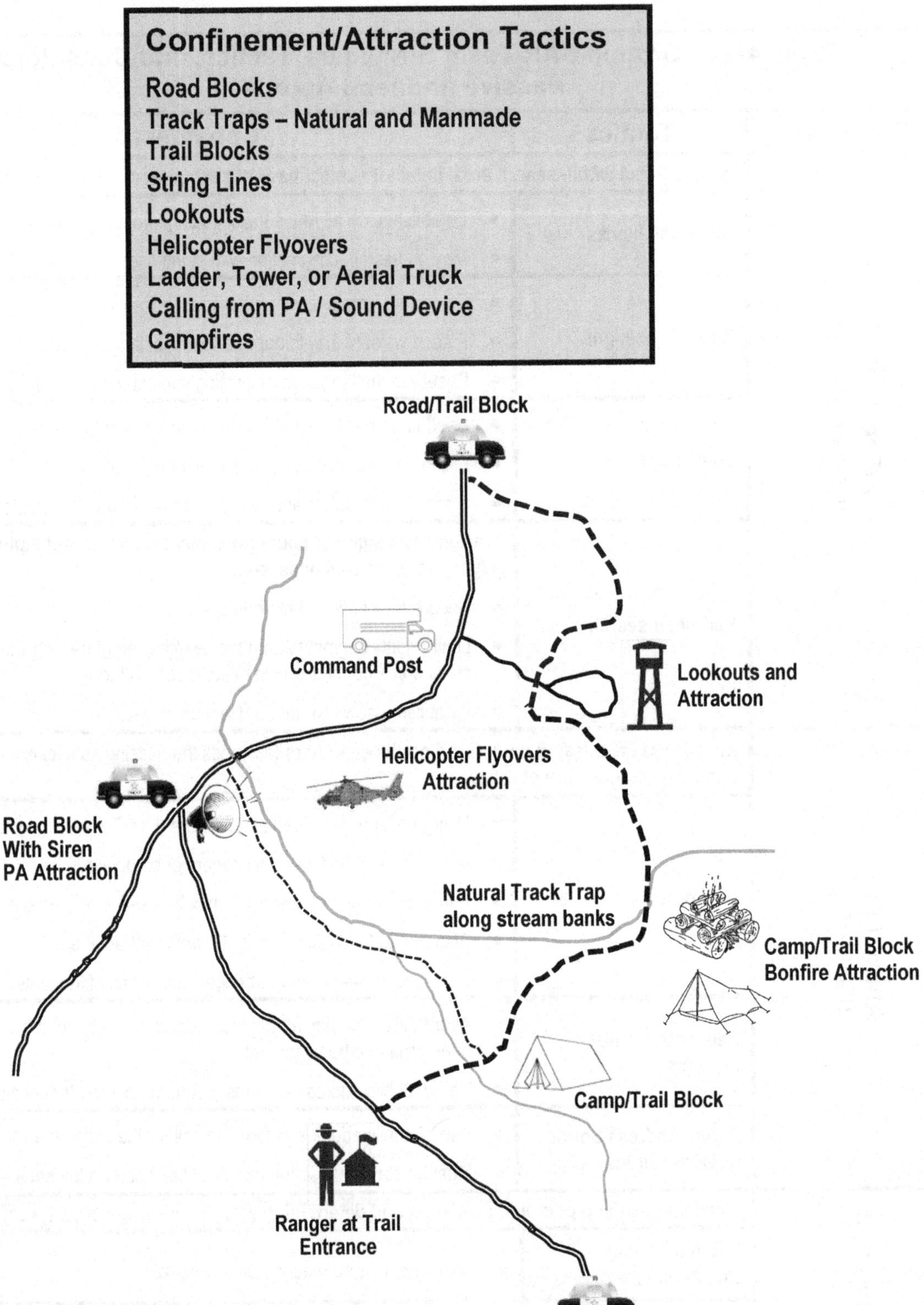

Figure 4-4-1: Confinement/Attraction Tactics

Table 4-4-2: Ground SAR Team Strategies, Tactics, and Considerations Active (Direct) Approach		
Strategy	**Tactics**	**Considerations**
Rapid (Hasty) Search	**Checking the immediate area, trails, roads, buildings, campsites, and specific areas of high probability.** Rapid searches are normally conducted by the initially responding officer and conducted in the immediate area of where the missing subject went missing.	
	Linear Features Search (Trackline)	**Determine most likely route subject would have gone and quickly cover this route.** • Often used for lost hikers, walkers; • Conducted by initial responders; • Need to specify distance from feature for desired coverage; • Follows travel aids (trails, drainages, etc.); • Navigation is the greatest challenge, start task at clear point; • Downhill preferable; • Clue awareness, especially at decision points, is critical; and • Easily combined with sound and tracking techniques.
	Points of Interest (Spot)	**Thoroughly cover a specific area in which the subject may be located** • Check Scenic overlooks, bathrooms, playgrounds, swim pools, bodies of water; and • Check residence, outbuildings, and possible locations the missing subject may have visited.

Table 4-4-2 is continued on the next page.

Table 4-4-2 (continued)

<table>
<tr><th>Strategy</th><th>Tactics</th><th>Considerations</th></tr>
<tr><td rowspan="3">Segment (Area) Search</td><td colspan="2">Requires easily identifiable boundaries for search teams. Often, these boundaries are also likely routes the missing subject may have traveled, Topographical features natural (ridges, streams, drainages, field edges) or manmade (roads, trails, fences, power or utility transmission line clearings).
• Tactics depend on type of resources available and terrain.</td></tr>
<tr><td>Route (Area) Search (Figure 4-4-2 and Figure 4-4-3)</td><td>Systematic search in which team members follow tracks parallel to a side boundary & maintain a predetermined separation.
• Search area may be covered in one or more passes;
• All searchers should attempt to walk in nearly straight lines parallel to the edge of the area, providing uniform, predictable coverage of the entire area;
• The base line is usually formed along a search area boundary with searchers properly spaced apart;
o Spacing will determine if the Tactic is less thorough or thorough
o Spacing will be determined by the density of vegetation in the area
• Purposeful wandering may be employed;
• Search leaders should select search area boundaries that are easy for the search teams to recognize and follow; may be natural or man-made, pre-existing or set up by the search teams; and
• Team Leader keeps the team moving in the right direction, at a reasonable pace, and maintaining proper searcher separation.</td></tr>
<tr><td>Area (Grid) Search (Figure 4-4-4 and Figure 4-4-5))</td><td>Thorough tactic to raise POD and look for unresponsive subjects.
• Competent flankers required, if using emergent volunteers more experienced crew leaders required;
• The base line is usually formed along a search area boundary with searchers properly spaced apart;
o Spacing tight to ensure thorough tactical search
o Spacing will be determined by the density of vegetation in the area
• Evidence type search may be shoulder to shoulder and conducted on hands and knees;
• Direction of the search follows a specific compass bearing;
• Purposeful wandering may be employed;
• Instead of large lines, used staggered starts and flagging tape; and
• Tight Grid Searching (thorough) is manpower intensive and should be consider as a last resort after other search tactics have been used and narrowed the search area down.</td></tr>
</table>

Table 4-4-2 is continued on the next page.

Table 4-4-2 (continued)

Strategy	Tactics	Considerations
Segment (Area) Search	Sound Sweep	**A method to search a relatively large area quickly. It requires that the missing person be responsive.** • Must be carefully coordinated, audible, and requires all to pause and listen for response; • Sounds can often be heard even when the missing person cannot be seen; and • Searchers will use whistles or call the missing subjects name at specific intervals.
	Expanding Circle Search (Figure 4-4-6)	**Trackers search for clues and sign expanding outward in a spiral from a defined starting point (e.g., LKP, PLS, previous clue).** • Tactic only effective for small areas; • Typically used at IPP or where a clue is found as a follow-up tactic; • Best with experienced tracking team to avoid destroying clues/tracks; and • May be used following the contour of a hilltop working downward.

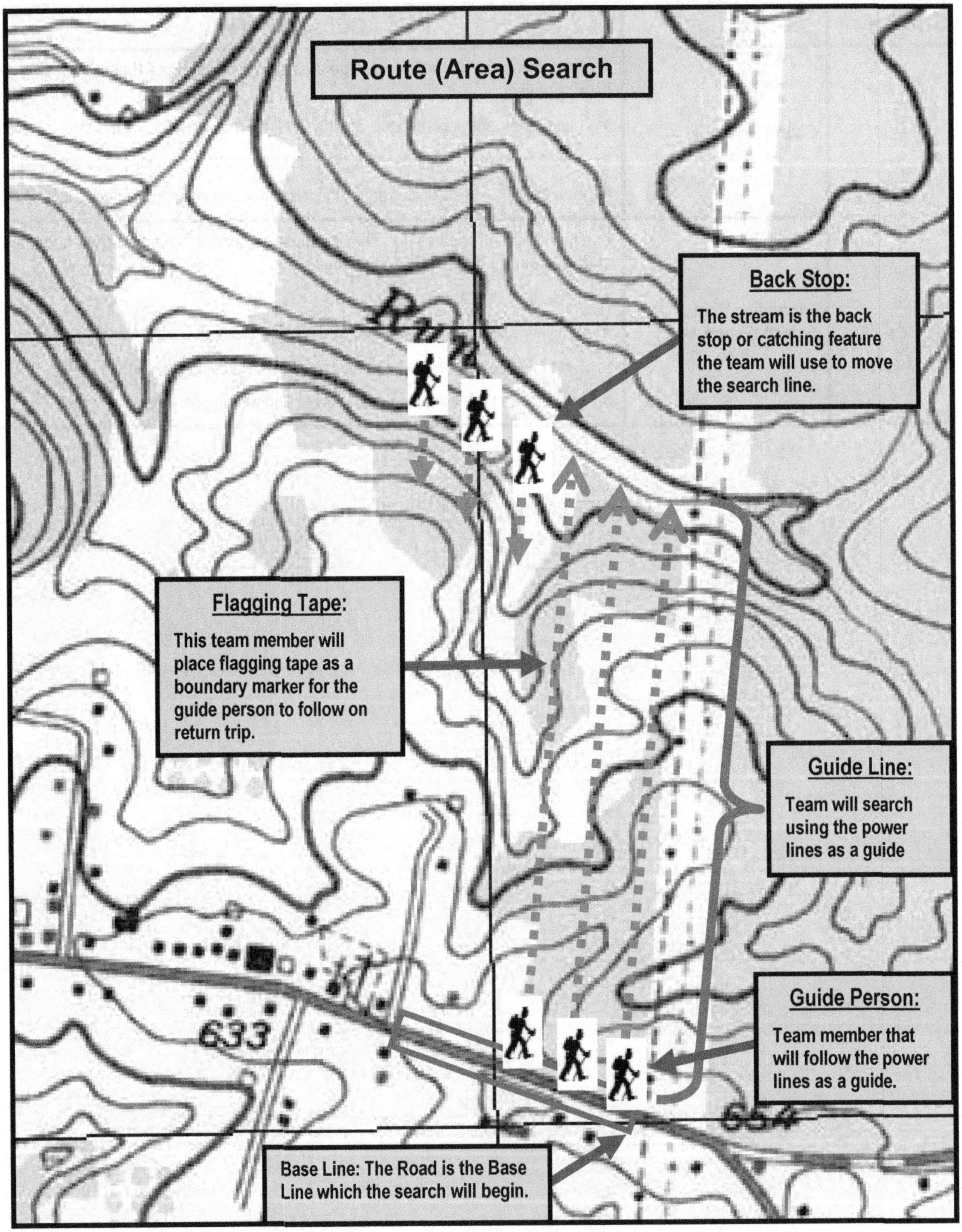

Figure 4-4-2: Route (Area) Search

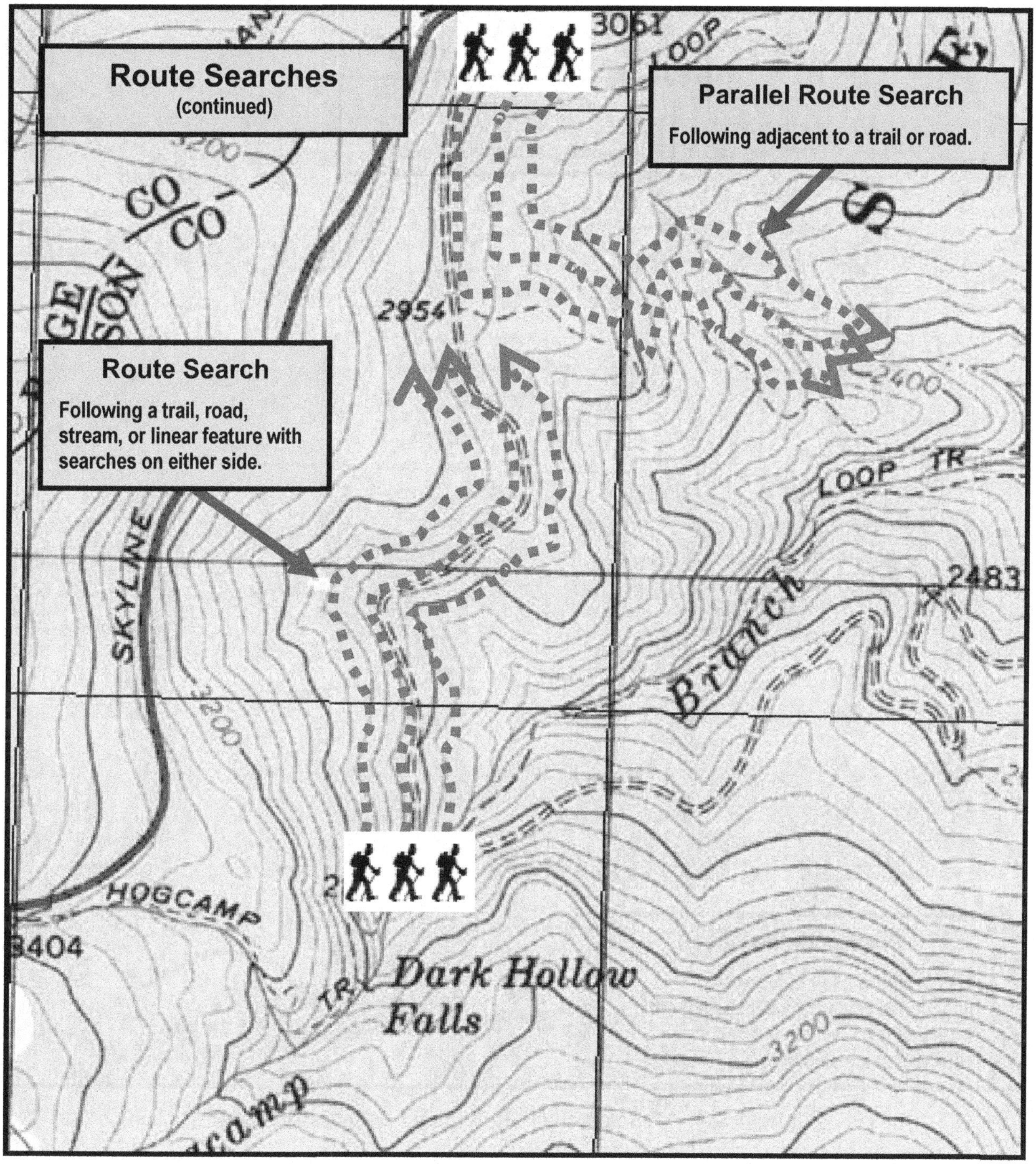

Figure 4-4-3: Route Searches (continued)

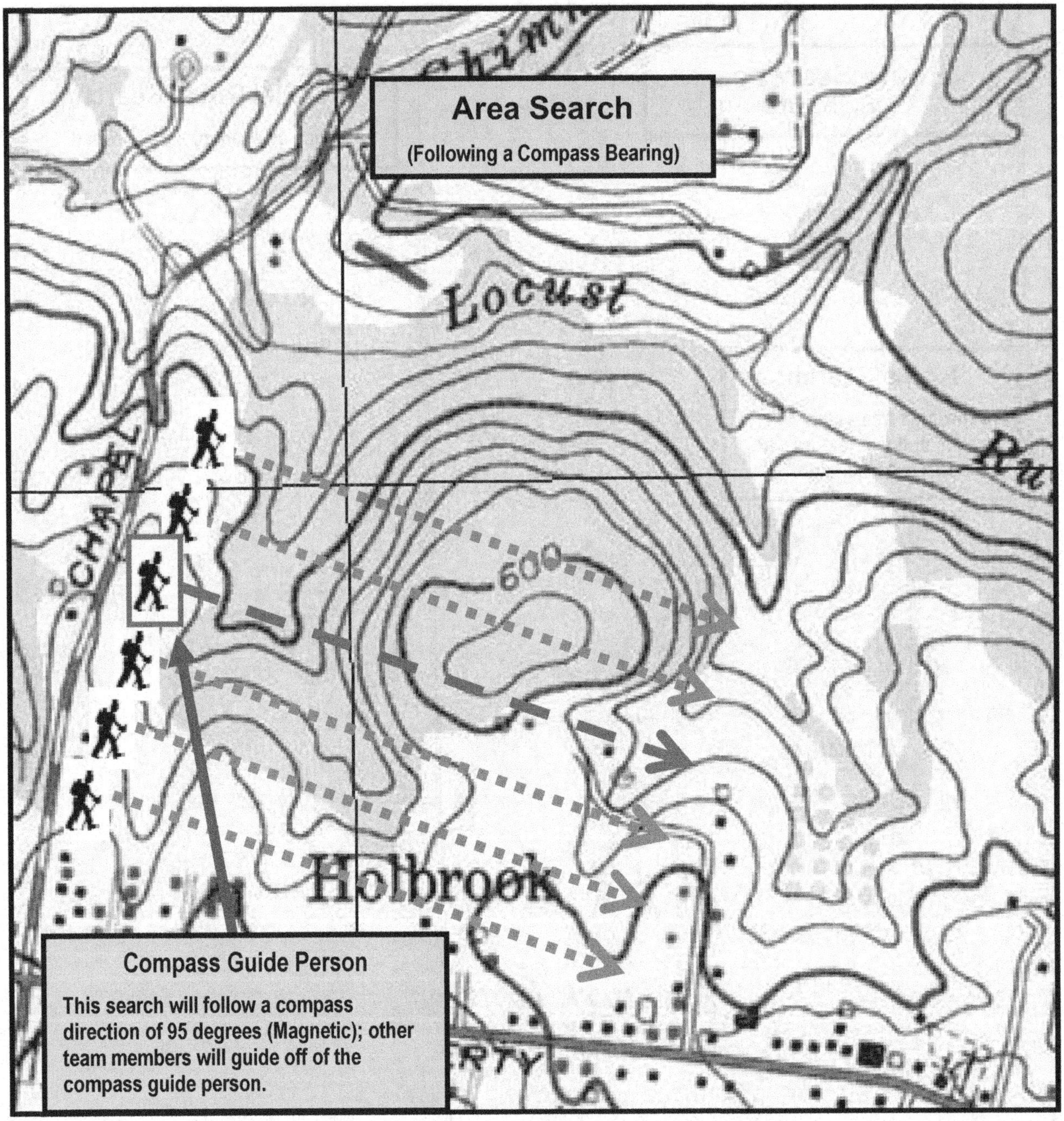

Figure 4-4-4: Area Search (Following a Compass Bearing)

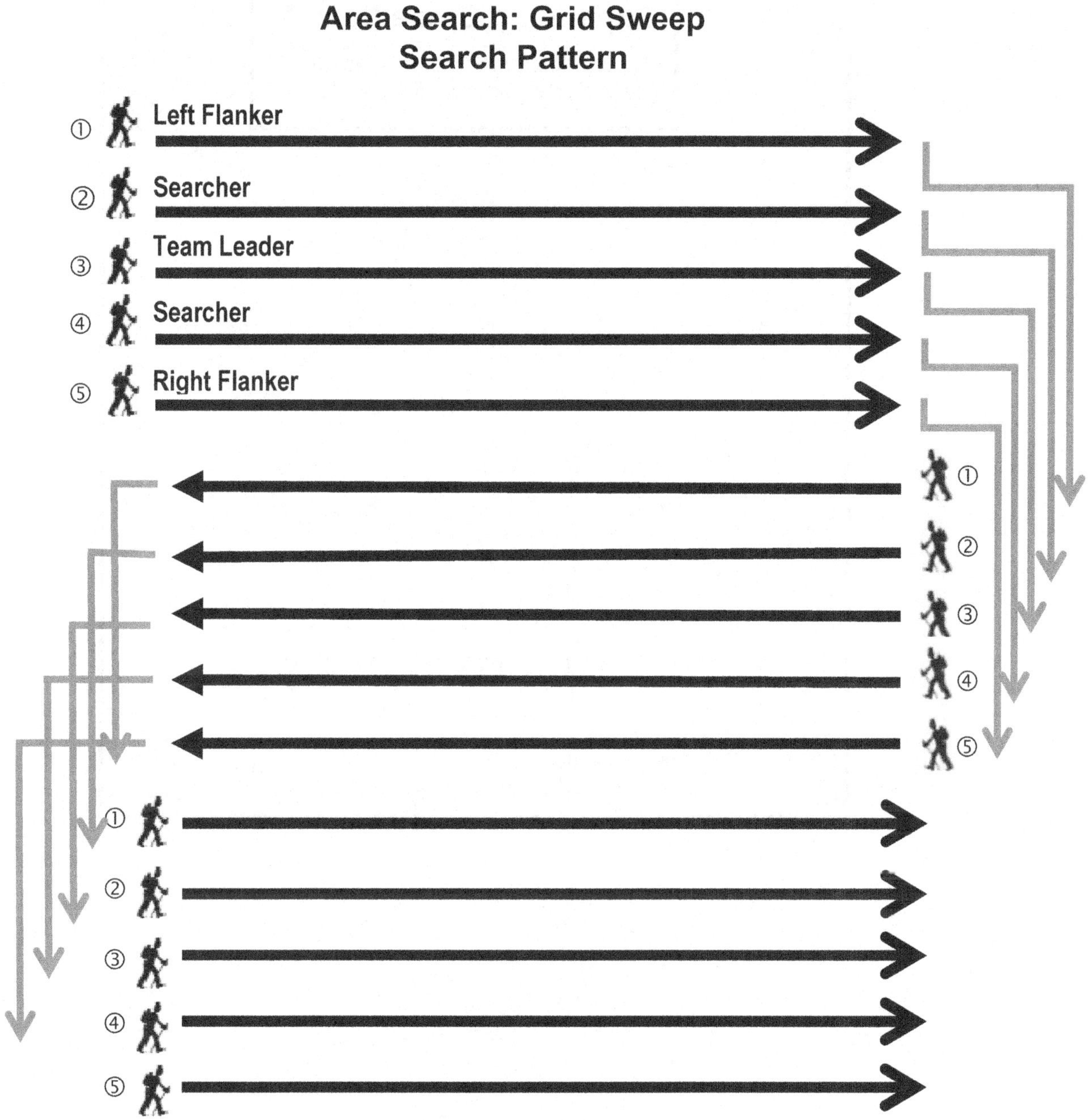

Figure 4-4-5: Area Search: Grid Sweep Search Pattern

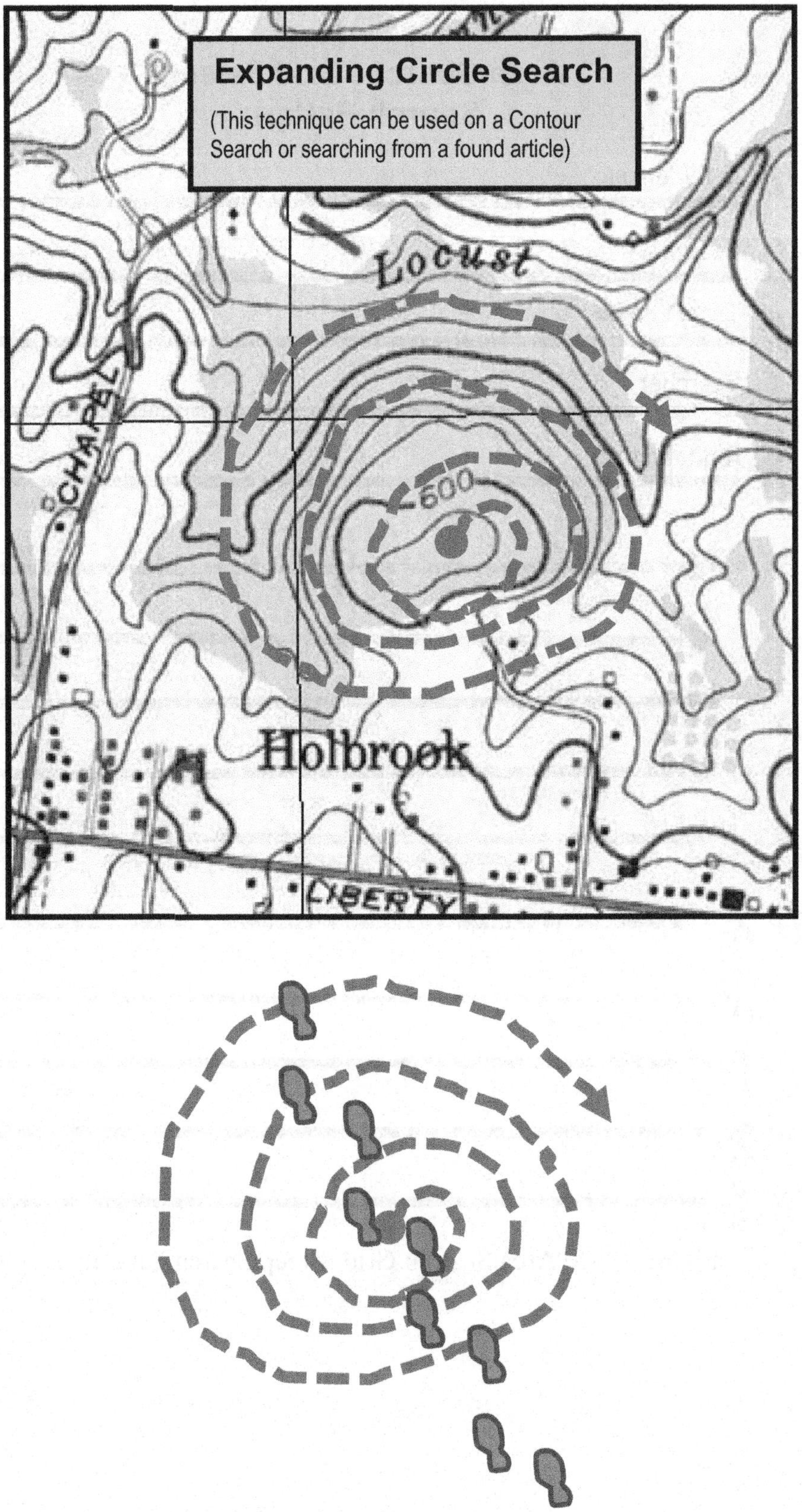

Figure 4-4-6: Expanding Circle Search

Canine SAR Teams

Table 4-4-3: Canine SAR Team Tactics and Considerations	
Tactics	**Considerations**
Coordinated Dog/ Tracker	**Combine advantages of air-scent dog and man trackers into one task.** • Able to deploy man tracker and air-scent dog simultaneously into high probability area; • Tracker able to pick up tracks at decision points and track traps; and • Need to train on tactic prior to incident.
Expanding Circle	**Thorough check around IPP/PLS/LKP to locate clues or the subject.** • Common at the IPP and around clues; and • Useful in searching for persons with dementia, despondent, abductions, and children.
Trail	**Cover high probability areas, quickly using minimal resources.** • Same as ground team with extra capability of canine.
Segment	**Thorough technique (wind permitting) to cover large area with minimal resources.** • Canine segments the same as ground team segments; • Establish boundaries for segment; consult handlers for optimal size; and • Handler will determine method to work segment depending upon wind conditions.
Corridor	**Quickly cover the highest probability area with minimal resources.** • Team will work trail/road and a specified distance from the feature; • Best for subjects often located short distances from linear features; and • Need to specify distance from feature for desired coverage.
IPP/PLS/LKP	**Validate LKP/PLS/IPP, determine direction of travel, find subject's trail, locate subject.** • Important to know if others have been present at IPP and previous search efforts; • If a trail is located, best to validate with multiple dogs; • Allow dog to go where it wants to go. Handler should be coupled with a team member with navigation skills; and • Special needs in urban setting.
Specific Location	**Verify clues and sightings, locate alternative LKPs, determine direction of travel, locate subject:** • Start dog from the site of a clue or sighting; and • If scent-discriminating adds validity to clue when dog follows trail.
Containment	**Determine if the subject crossed a feature.** • Effective if performed with a trailing dog; • Not all trailing dogs are able to perform; • Limited to a small area; and • Trailing dogs are not able to maintain focus when not following a scent.

Mounted, All Terrain Vehicle (ATV), Snowmobile, and Bicycle SAR Teams

Table 4-4-4: Mounted, ATV, Snowmobile, Bicycle SAR Team Considerations	
Type	**Considerations**
Mounted (horse)	• Mounted teams can be tasked the same as ground teams; • Experienced team can navigate difficult trails; • Offers a higher search platform, increasing detection range; • Horse may give an alert if it detects something out of the ordinary; • Terrain may preclude deploying horses; • Consider mounted teams for corridor and segment searching; and • An experienced rider may or may not mean experience off-trail.
ATV (quad)	• Covers trails quickly and can carry additional gear; • Clue awareness diminished and the ability to hear is limited; • Can destroy clues; • Riders should get off the ATV frequently to check potential track traps and decision points; and • Move away or turn off the machine to listen for sound clues.
Snowmobile	• Refer to ATV considerations; and • Ability to travel cross-country will depend upon vegetation and amount of snow cover.
Bicycle	• Advantages: hear better, erase fewer clues, and easier to stop/examine track traps and decision points; • Disadvantages: less capable of traveling off-trail and is limited in the amount of gear that can be carried; and • Highly effective in an urban setting.

Tracker SAR Teams

Table 4-4-5: Tracker SAR Team Tactics and Considerations	
Overall strategy: Location of signs, determine direction of travel, follow tracks, and normally work in teams of 2-3 persons.	
Tactics	**Considerations**
Step by Step	**Thorough technique to follow tracks.** • 2-3 team members; • Most basic tactic is to follow tracks; generally slow, but confirms tracks are all tied together; and • Initial tactic taught to most beginning trackers.
Binary sign cutting	**Determine if subject's tracks are in a specific area.**
Linear Features Trackline	**Follow feature to determine if subject is on a specific trail, turned off the trail, or turned onto a different trail.** • Starting point typically road, trail, railroad track, creek, gully, or other linear feature; and • Team follows linear feature, special emphasis on decision points and natural track-traps.

Aeronautical SAR Resources

Table 4-4-6: Aeronautical SAR Resource Strategies, Tactics, and Considerations		
Strategy	**Tactic**	**Considerations**
Rapid (Hasty) Search (Aviation)	Route Search	**Search along subject's intended route(s).** • Follow intended route from point of departure to intended destination; • May also follow alternative routes, roads, or trails; • Multiple passes easily achieved; consider flying at different times of day in different directions; and • Route searches usually expanded to trackline pattern
	Trackline Search (Figure 4-4-7)	**Rapid search with multiple passes of intended route.** • Tactic starts with route search and then two patterns possible; • Single non-return; searches either side of the track line; and • Single unit-return, searches trackline in both directions combined with searching both sides of the trackline.
	Electronic	**Rapid search with electronic direction finding equipment able to detect distress beacons.** • Route search with DF equipment; • Range of detection varies with search unit altitude and/or terrain; and • Specific coordinate search for activated distress beacon.
Area Search (Aviation)	Parallel Search Pattern (Figure 4-4-8)	**Systematic search of a rectangular area with multiple passes.** • Straight search legs are usually aligned to cardinal headings parallel to long axis of search area; and • Pattern typically used for large, level search area when uniform coverage desired.
	Creeping Line Grid (Figure 4-4-9)	**Systematic search of a rectangular area that allows search at one end or adjustments for sun angle.** • Specialized type of parallel pattern where direction of creep is along major axis; • Used to cover one end of an area first; • Able to change search direction if sun glare would hinder searching; and • Used to provoke a signal if subject has signaling devices.
	Sector Search (Figure 4-4-10)	**Extensive coverage of center point at different angles and with multiple passes.** • Similar to expanding circle and square; • Used when search area is not extensive; • Concentrated over the central point; • Generally radius no greater than 20 miles; and • Both rotary and fixed wing capable.

Table 4-4-6 is continued on the next page.

Table 4-4-6 (continued)

	Expanding Square (Figure 4-4-11)	**Thorough check around center point with rectangular legs.** • Starting point typically similar to expanding circle; • Pattern uses straight legs with turns only at corners of each square; • May require continuous reprogramming of navigation units; and • Leg spacing 0.25-0.5 NM in missing person searches.
	Contour (Figure 4-4-12)	**Systematic search of mountainous terrain.** • Used in mountains and hilly terrain; • Only one aircraft should be assigned to area; • Aircraft starts on the up-wind at the top and works downhill, then returns to the top and works down-wind side; • Each contour circuit typically a drop of 500 ft in altitude; • Crew must be experienced in mountain terrain and should have detailed, large scale maps of area; • Weather conditions must be favorable; and • Aircraft needs high climb rate.
	Direction Finding (See Section 5-4: Distress Beacon Search Concepts and Figure 5-4-3)	**Airborne location of electronic distress signals** • Used when the equipment on-board has the ability to receive the distress beacon signal; and

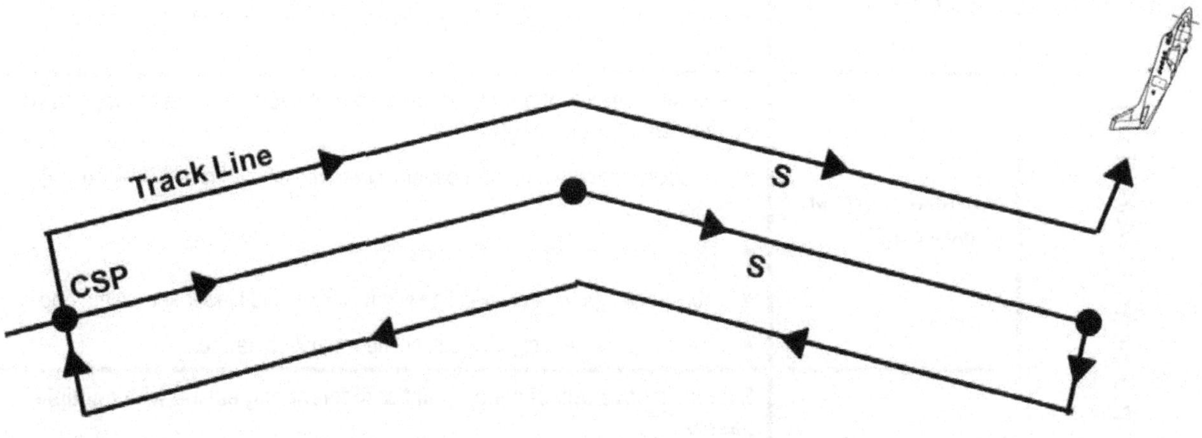

Figure 4-4-7: Aviation Trackline Search Pattern

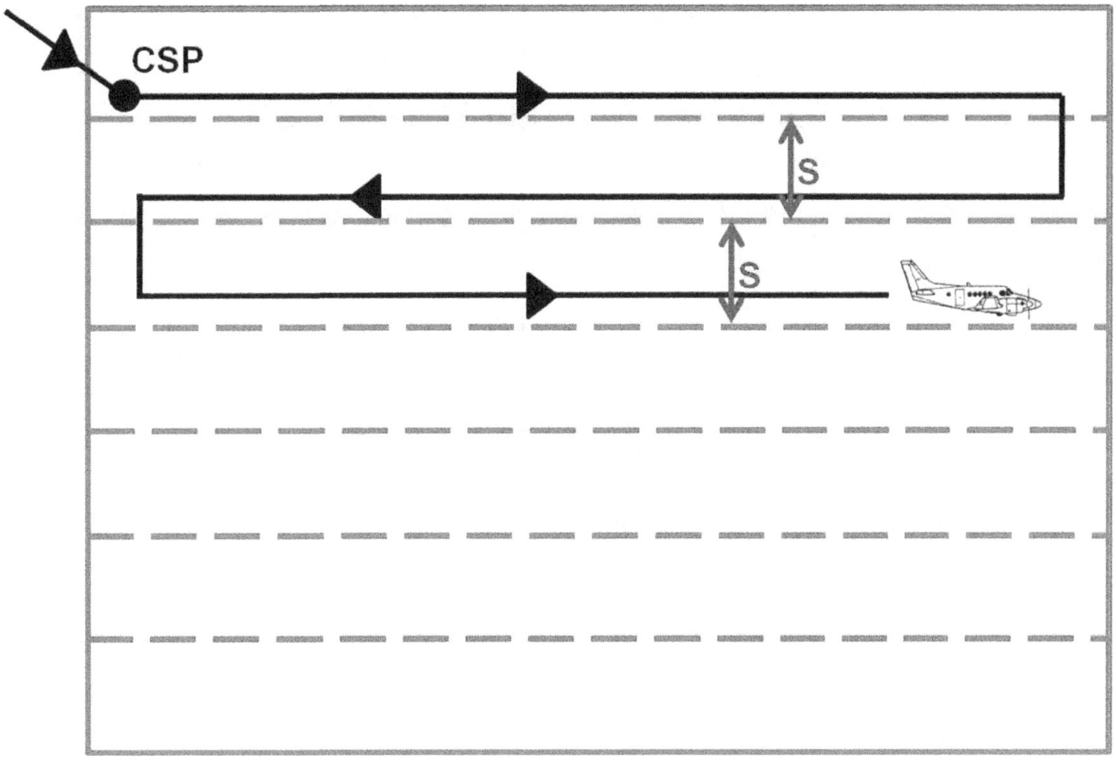

Figure 4-4-8: Aviation Parallel Search Pattern

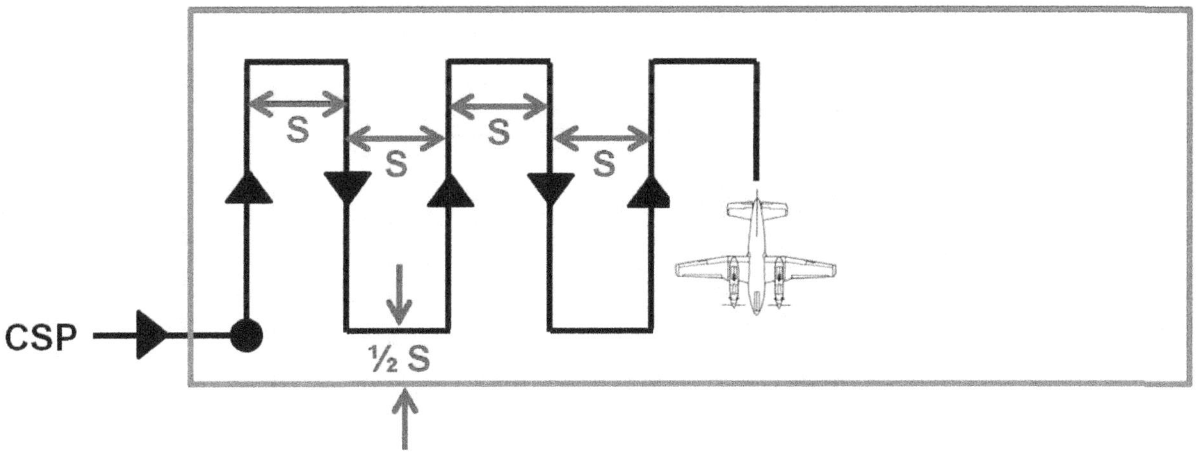

Figure 4-4-9: Aviation Creeping Line Grid Search

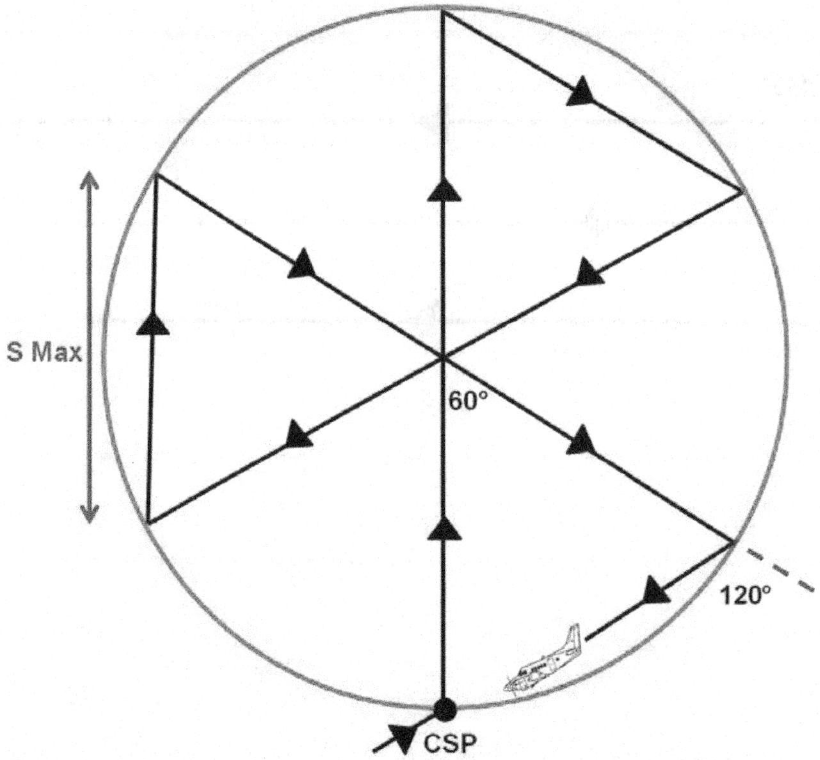

Figure 4-4-10: Aviation Sector Pattern Search

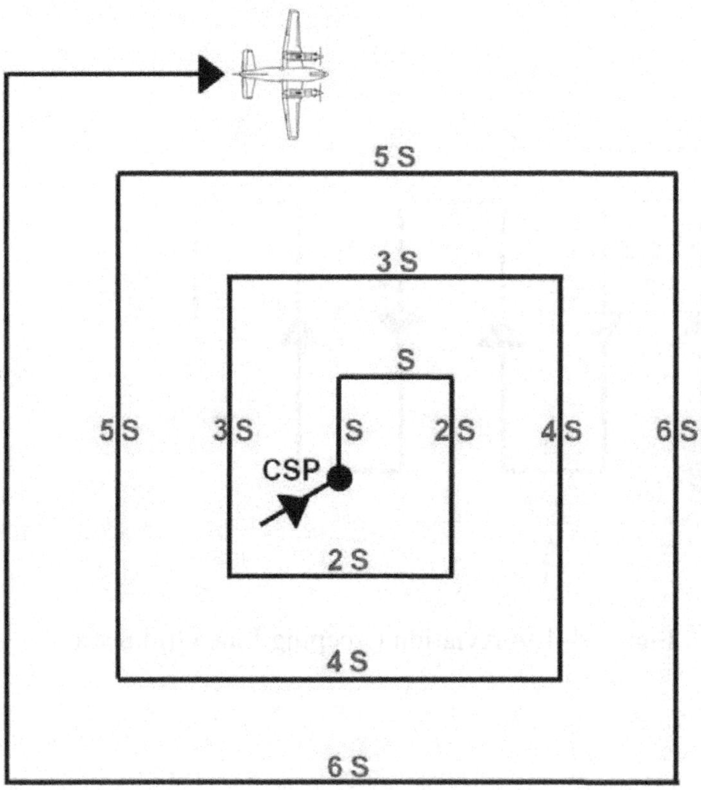

Figure 4-4-11: Aviation Expanding Square Pattern Search

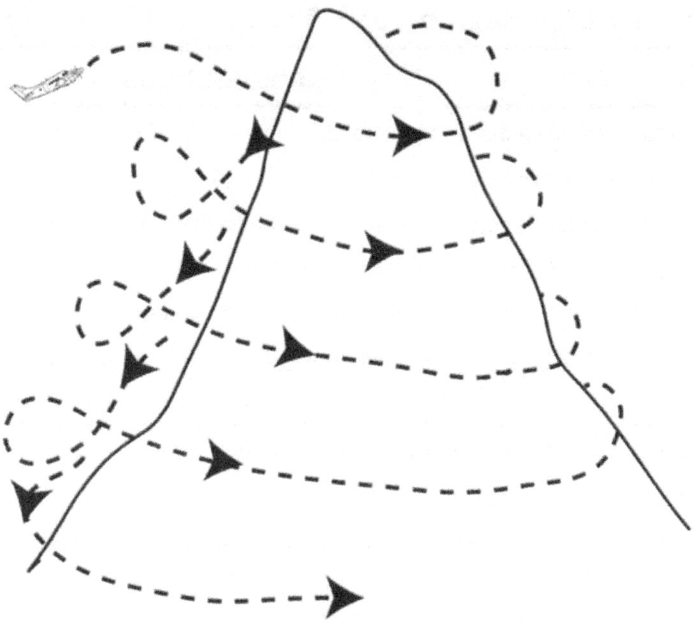

Figure 4-4-12: Aviation Contour Search Pattern

Boat Search Teams

Motorboat search tactics are similar to aircraft search tactics. While the patterns are the same, the spacing for searching from a small motorboat is considerably less, especially if looking for a person in the water. See the Coast Guard Addendum for additional information.

Dive Search Teams

Dive search tactics are rooted in land and air patterns, but limited visibility, chaotic bottom contours, and underwater hazards require the use of special patterns. While several options exist, the simpler the pattern that meets the objective, the better. Selection of search pattern will depend upon bottom type, depth, visibility, number of divers, and experience of divers, search object, and diving environment.

Table 4-4-7 on the next page provides an overview of dive search team tactics and considerations.

Table 4-4-7 Dive Search Team Tactics and Considerations	
Tactic	**Considerations**
Circular	**Systematic search around a fixed point.** • Involves one to several divers; • Anchor (ascend/descend line) used to mark start must remain fixed; • Line attached to anchor to guide team in circle; • Mark starting point of search so aware when 360° completed; and • When a 360° sweep is done, diver farthest from anchor stays and the rest of the team shifts further out.
Semi-circular	**Systematic semi-circular search from wall or pier.** • Similar to circular search; • Instead of a fixed anchor point, the fixed point is usually attached to pier piling or wall near the bottom; and • Pilings or wall used as starting and ending points.
Straight Sweep	**Cover a large area with multiple divers and good visibility.** • Used to cover rectangular search areas with multiple divers; • Anchor placed for guideline well beyond where search object may be located; • With taunt guideline second line attached with a loop at right angle; • Divers swim out to anchor, form line along second line; • Spacing determined by visibility; and • Requires teamwork and coordination.
Straight Sweep along shoreline	**Tactic effective for those likely to be found close to shore.** • Divers spaced along guideline as visibility will permit; • Guideline held by someone walking along shore keeping up with divers; and • If object not located, last diver holds position while others move further out on guideline.
Grid (Checkerboard)	**Used to systematically grid an area.** • Requires four anchors and four floats; • If free of obstructions use guidelines beneath the surface; • Divers swim assisted by the guidelines or use compass within the grid area; and • If the search is extensive, lay out pattern to last one tank of air.
Search with a weighted line	**Searching around oddly shaped objects or pilings.** • Guideline held by person on the pier; • Line heavily weighed for reference point; and • Divers must swim slowly to avoid getting ahead of the guideline.

Table 4-4-7 is continued on the next page.

Table 4-4-7 (continued)

Planning Board	**Search large area for a large search object.** • Planning board or sea sled deployed behind a boat; • Divers must prevent rapid ascents; • Diver needs a marking device if locate possible search targets; and • Boat operator must stay away from marker float.
Towing Divers	**Search large areas with unlimited visibility.** • Used for searching oceans or rivers where bottom visible; • Divers pulled behind boat without planning board; • Number of divers dependent upon current and speed of boat; and • Each diver should have own rope.
Compass Search	**Conduct area search when unable to use guidelines.** • Used when bottom conditions do not allow search lines (grass, kelp, snags, underwater hazards); • Compass used to maintain lines and avoid overlap and gaps; and • May be possible to use with underwater vehicle.
Jack-Stay or "Z" Pattern	**Cover an area with a steep embankment.** • Diver(s) search from deep depths to shallower depth; • Two down lines (with float) attached to 15-20 pound anchor; • Length of line (50-75 ft long) attached to anchors; • Diver searches along line from anchor to anchor. Upon reaching opposite anchor, the anchor moved the predetermined sweep spacing; and • Diver reverses direction, with some overlap, and repeats sequence.
Quick Search	**Used as initial tactic for subjects close to shore or pier.** • Requires two divers; • First diver enters water (at point directed by witnesses) with 100 feet of line held by second diver (tender) at surface; • Diver submerges to the bottom, and swims until line pulled tight; and • Diver then searches in an arc, with the tender keeping the line taunt as the diver search back and forth with a decreasing radius.
Snag Search	**Rapid search performed by one diver to locate a large object.** • Diver secures a line at point assumed object went into water; • Diver swims past point object should be located with line and descends; • Diver swims back to shore/pier in an arc; • Large objects in the path of the line should be snagged; and • Diver follows line back to snagged object.

This page intentionally left blank.

Section 4-5: Rescue

Rescue Operations

Rescue by Aircraft

Specialized Environments

Care of Rescued Personnel

Rescue Operations

When the subject has been located, the next step is to perform rescue operations. The SMC/IC, OSC, searchers, pilot in command, etc., must decide the best method of rescue The following factors may apply:

- Action taken by the on-scene resources;
- Location and disposition of the subject;
- Subject's medical condition;
- Environmental and weather conditions, observed and forecasted;
- Available SAR resources; and
- Risks to rescuing personnel.

In particular, the SMC/IC must carefully consider the risks and challenges associated with helicopter extraction and transport. The CISAR Addendum has additional information concerning management of aircraft operations.

Typically, rescue operations involve four phases as described in Table 4-5-1 below:

Table 4-5-1: Four Phases of Search and Rescue

Phase	Description
Locate	The search effort.
Access	Requires direct contact with the missing subject. It could be as easy as walking to the person or may involve specialized equipment or resources.
Stabilize	Caring for injuries as well as providing for physical comfort and safety of the subject and any others involved in the incident.
Transport	Transport the subject to a place of safety or appropriate medical facility. Determine if specialized resources will be required or if transport can be conducted by personnel on scene. Decide on the easiest route.

Specialized Environments

Rescues occur throughout the U.S. in a wide variety of complex environments. The SMC/IC in charge of such a rescue, whether at the end of a long search, or responding directly to a person in distress, should realize that most incidents are resolved by well-trained responders.

Some of the specialized environments that an SMC/IC may encounter are summarized in Table 4-5-2 in the next page:

Table 4-5-2: Specialized Environments Encountered in Rescue Operations	
Environment	**Description**
Mountain	Long steep slopes, vertical snow and ice, vertical rock faces, avalanches, glaciers and crevasses.
Water	Lakes, frozen surfaces (thin ice), whitewater streams and rivers, flashfloods, slowly rising floods.
Restricted Access	Caves, mines air shafts, abandoned wells.
Urban/Disaster	Building collapse and other special urban/disaster rescue conditions
Weather	High winds, snow, blizzards, excessive heat.
Hazardous Materials	Chemicals and other dangerous materials that may impact rescuers and victims.
Natural Obstructions	Challenges concerning tall trees, other heavy vegetation, rocky terrain, etc.

Aircraft Rescue

In some cases aircraft may be used for rescue operations. Each aircraft has operational and technical limitations and should not be used for operations which are not suitable. When possible, a rescue operation by aircraft should be coordinated with surface resources.

Fixed wing aircraft can be used to:

- Deploy equipment to persons in distress;
- Direct ground SAR resources to the persons in distress; and
- Mark the position of the person(s) in distress by serving as a radio and radar beacon, showing lights, dropping flares, and providing radio signals for direction finding and homing by ground resources.

More often, helicopters are a more suitable air resource to use in rescue operations. The helicopter's unique flying capabilities allow them to be used in smaller, hard to reach areas. However there are some special concerns to be aware of when using a helicopter:

- Operations on the ground may be hampered by the noise and rotor wash.
- Close coordination between the helicopter and ground resource is required (communication using international hand signals may be appropriate);
- Helicopters generally are limited on the number of persons and amount of equipment they can carry;
- Helicopters are susceptible to icing and have limited fuel reserves (it may be necessary to dispatch fixed wing aircraft to check weather enroute to ensure the helicopter can safely transit to the subject);
- Rescues by helicopter landing creates additional concerns such as:
  - Turbulence;
  - Unlevel terrain;
  - Clearing of loose debris;

- Altitude limitations (high altitudes reduce helicopter performance greatly); and
- Identifying safe landing and takeoff paths.

- For the evacuation of persons, helicopter may be equipped with the rescue sling, basket, net, litter, or seat.

Care of Rescued Personnel

The SMC/IC must ensure that the subject is transported to a *place of safety*. The SMC/IC should anticipate that the subject may require medical treatment and should consider having an ambulance and hospital facilities alerted and/or nearby for immediate transport.

SAR personnel should never leave the rescued persons alone, particularly if they are injured or showing signs of shock, hypothermia, physical, or mental exhaustion.

When selecting the method of transportation to medical facilities, consider the following:

- Condition of the person(s) rescued;
- Capability of the rescue resource to reach the distressed persons in the shortest possible time;
- Medical training, qualifications, and operational capabilities of the SAR personnel; and
- Difficulties that may be encountered by SAR personnel.

This page intentionally left blank.

Section 4-6: Non-Rescue Operations

Saving Property

Human Remains Recovery

Contact with Human Remains

Saving Property

The primary emphasis of SAR operations is to assist persons in distress. Land and aeronautical SAR missions should not normally be performed for the purpose of preventing or mitigating property loss or damage or for salvage or recovery of property when those actions are not essential to the saving of persons in distress.

Beneficial secondary consequences of a rescue operation may be to prevent environmental damage or remove hazards, but these are not normally considered part of the rescue operation's objective.

When faced with "saving" or "salving" property, there are several factors to consider:

- Cost and risk to personnel and equipment involved;
- Whether abandoning property may introduce other problems;
- Whether proper facilities, equipment, and skill are available to conduct operations; and
- Advice of personnel on-scene who are in the best position to assess the situation.

When saving or salving property, it can be a natural extension of lifesaving efforts or a means of lifesaving. However, saving property should never jeopardize persons or equipment.

Human Remains Recovery

Disasters have demonstrated that the risk of epidemic disease transmission from human remains is negligible. Unless the affected population was already experiencing a disease suitable for epidemic development, the event should not create such a situation. Most victims die from traumatic events and not from pre-existing disease.

Disease transmission requires a:

1. Contagious agent;
2. Method of transmission; and
3. Susceptible population to infect.

Typical pathogens in the human body normally die off when the host dies, although not immediately. The risk of transmission is no greater than that for routine handling of human remains.

Water supplies contaminated with decaying human remains can serve as a method of transmission of illnesses, particularly gastroenteritis, but a non-breathing body presents minimal transmissibility.

Contact with Human Remains

Human remains may contain blood-borne viruses such as hepatitis B and C, HIV, and bacteria that causes diarrheal diseases, such as shigella and salmonella. These viruses and bacteria do not pose a risk to someone walking nearby, nor do they cause significant environmental contamination.

In flood water, bacteria and viruses from human remains are a minor part of the overall contamination that can include uncontrolled sewage, a variety of soil and water organisms, and household and industrial chemicals. There are no additional practices or precautions for flood water related to human remains, beyond what is normally required for safe food and drinking water, standard hygiene, and first aid.

However, for people who must directly handle remains such as responders in the course of conducting SAR operations and recovery personnel, there can be a risk of exposure to such viruses or bacteria.

SAR responders who must handle human remains should consider the following precautions:

- Protect the face from splashes of body fluids and fecal material;
    - Use a plastic face-shield or a combination of eye protection (indirectly vented safety goggles are a good choice if available; safety glasses will only provide limited protection) and a surgical mask; and
    - In extreme situations, a cloth tied over the nose and mouth can be used to block splashes;
- Protect the hands from direct contact with body fluids, and also from cuts, puncture wounds, or other injuries that break the skin that might be caused by sharp environmental debris or bone fragments;
    - A combination of cut proof inner layer glove and a latex or similar outer layer is preferable; and
    - Footwear should similarly protect against sharp debris.
- Clothing:
    - Disposable clothing should be made available and is recommended for human remains recovery (however, in some cases, traditional fabrics may be preferable owing to their strength - especially when lifting bodies);
    - Gowns or aprons should be worn during procedures that are likely to generate splashes of blood or other body fluids;
    - Closed, boot-style shoes are recommended (Wear rubber boots or appropriate shoe covers where there is potential for footwear to become grossly contaminated); and
    - Rain gear may also be useful in case of storms.
- Maintain hand hygiene to prevent transmission of diarrheal and other diseases from fecal materials on the hands.
- Wash hands with soap and water or with an alcohol-based hand cleaner immediately after removing the gloves.
- Give prompt care – including immediate cleansing with soap and clean water and a tetanus booster if indicated, to any wounds sustained during work with human remains.
- In addition to guarding physical safety, participate in any available programs to provide psychological and emotional support.

Section 4-7: Geo-referencing

Introduction

In the aftermath of Hurricane Katrina, the review of the federal, military, State, and local SAR response found that SAR agencies used different methods to communicate geographic information. This added confusion and complexity to an extremely large scale SAR operation.

Federal, State, tribal, territorial, local, and volunteer SAR responders working together face numerous challenges, including those relating to a lack of geospatial awareness. Three issues were identified during the Hurricane Katrina response:

- How do responders navigate when landmarks such as street signs and homes are blown away?
- How do responders communicate position in a common language?
- The final problem is resource de-confliction - the ability to ensure multiple assets are not inappropriately operating in the same area.

Resource de-confliction is a matter of safety, particularly with aircraft, to ensure the likelihood of a mid-air collision is minimized. Additionally, resource de-confliction is a matter of efficient and effective use of limited resources so that all areas receive an appropriate response.

In addition, geospatial communications and awareness are necessary to carry out search action plans (SAPs) and rescue action plans (RAPs).

What is Geo-referencing?

To geo-reference is to define location in physical space, and is crucial to making aerial and satellite imagery useful for mapping. Geo-referencing explains how position data (e.g., Global Positioning System (GPS) locations) relate to imagery and to a physical location.

Different maps may use different projection systems. Geo-referencing tools contain methods to combine and overlay these maps with minimum distortion.

Using geo-referencing methods, data obtained from observation or surveying may be given a point of reference from topographic maps already available.

Geo-Referencing Methods

Three geo-referencing methods are to be used for SAR operations anywhere in the United States, as indicated in the National SAR Committee geo-referencing matrix located in Table 5-7-1. Two point reference systems (USNG, Latitude/ Longitude) and one area reference system (GARS).

U.S. National Grid (USNG)

USNG is intended to create a more interoperable environment for developing location-based services within the United States and to increase the interoperability of location services appliances with printed map products by establishing a preferred nationally-consistent grid reference system. The USNG can be extended for use world-wide as a universal grid reference system, and can be easily plotted on USGS topographic maps by using a simple "read right, then up" method.

Note: USNG and the Military Grid Reference System (MGRS) are functionally equivalent when referenced to NAD 83 or WGS 84 datums.

The coordinates are easily translated to distance, as they are actually in meters. Thus the distance between two coordinates can quickly be determined in the field.

Figure 4-7-1 is an example of the USNG.

Figure 4-7-2 is an explanation for using the USNG.

Maps, Charts, and Coordinates

No single map/chart projection or coordinate/grid system will be perfect for all applications. In the case of projecting the earth's curved surface on a flat surface, distortion of one or more features will occur. The conventions for locating points on the earth's surface for purposes of nautical and aeronautical navigation (long distances on small scale charts) is generally best conducted using latitude and longitude (spherical coordinates). Locating points on large-scale maps and for ground navigation is generally best accomplished with Cartesian-style plane coordinates (i.e., U.S. National Grid (USNG)).

Large scale maps can treat the Earth's surface as a plane, rather than a complex sphere. Properly constructed large-scale maps (e.g., topographic maps) take curvature of the Earth into account.

Simple linear increments (i.e., meters) of plane coordinates are significantly easier for large-scale map users to handle accurately at high precision in the field than the more complex angular increments of latitude and longitude (i.e., degrees).

Additionally, there is a difference between a point reference system (i.e. USNG or Lat/Long) and global area reference system (GARS). A point reference system is used to identify a specific point to send SAR resources for a specific reason while a area reference system identifies a block area to send SAR resources for no specific response other than be available for a call if needed in that area.

USNG 100,000 Meter Grid with GZP Zone Designator

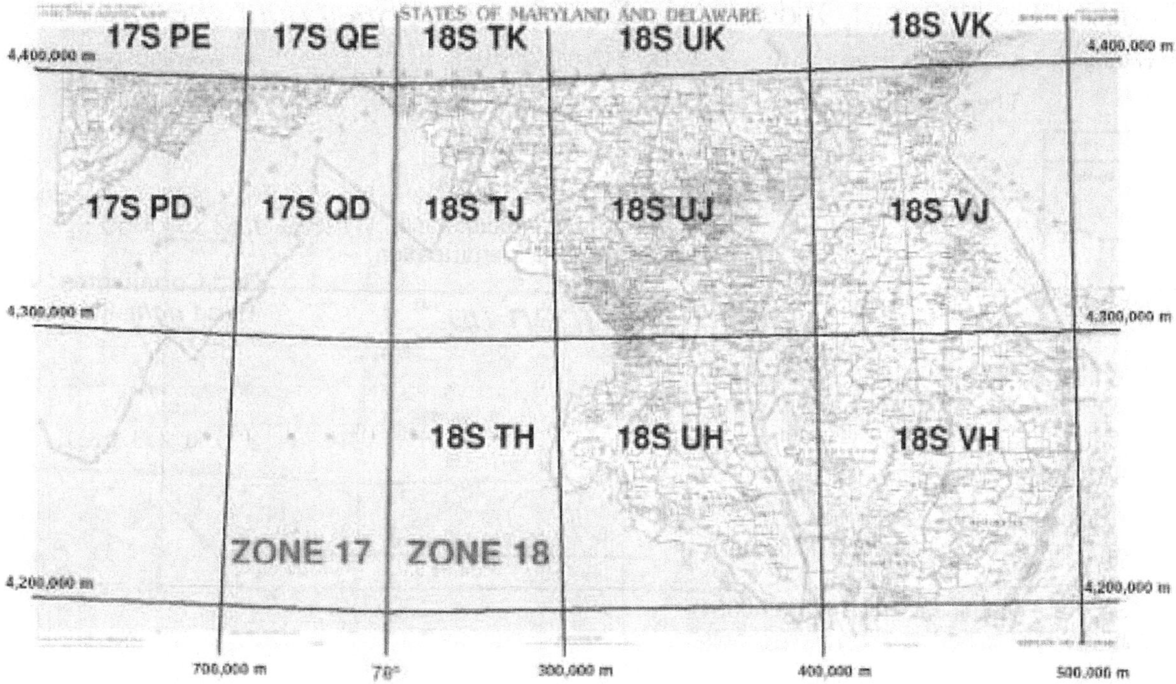

Figure 4-7-1: U.S. National Grid (USNG)

Figure 4-7-2: How to Use the USNG

Reading US National Grid (USNG) Coordinates: *"Read right, then up."*

Information Sheet 2/2 in this series. FGDC-STD-011-2001 From www.fgdc.gov/usng

The example below locates the Jefferson Pier at USNG: 18S UJ 23371 06519.

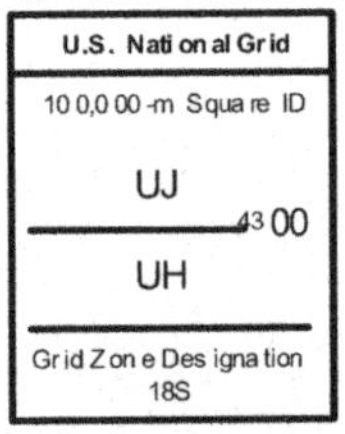

A USNG value has three components.

Some maps may give this leading information in a grid reference box.

Grid Zone Designation (GZD): 6° x 8° longitude zone / latitude band.
100,000-m Square Identification:

18S UJ 2337 0651

Grid Coordinates: Read ***right***, then ***up***.

"Read right, then up."

- Grid lines are identified by Principal Digits. Ignore the small superscript numbers like those in the lower left corner of this map.

Reading USNG Grid Coordinates.

- Coordinates are always given as an even number of digits (i.e. 23370651).

- Separate coordinates in half (2337 0651) into the easting and northing components.

- [1] Read right to grid line 23. [2] Then measure right another 370 meters. (Think 23.37)

- [3] Read up to grid line 06. [4] Then measure up another 510 meters. (Think 06.51)

Grid:	Point of Interest:
22 80 58	FDR Memorial:
23 10 54	George Mason Memorial:
2338 0710	Zero Milestone:
2275 0628	DC War Memorial:
22 20 65	Lincoln Memorial:

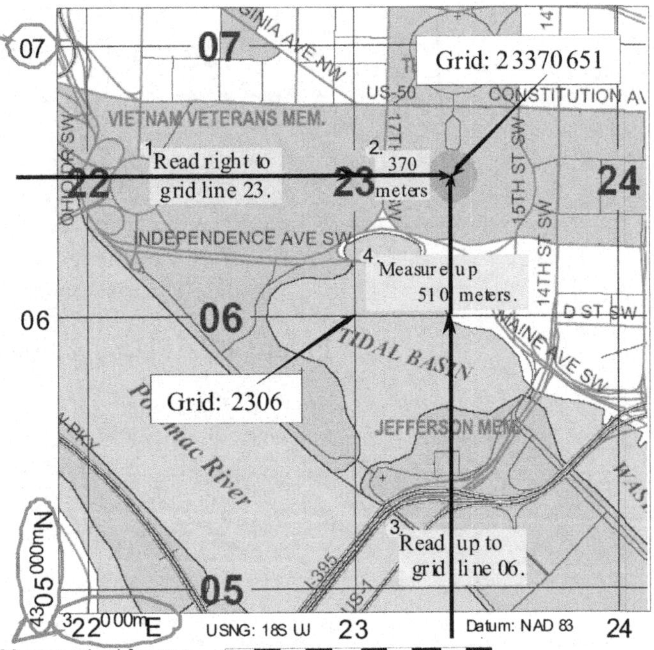

UTM numerical format.

Ignore the small UTM superscript numbers that are provided for reference purposes. UTM numerical values are best suited for determining direction and distance as in surveying. USNG alpha-numeric values are best suited for position referencing because they can be given as only grid coordinates in a local area and with only the required precision for a particular task.

Users determine the required precision. These values represent a point position (southwest corner) for an area of refinement.

Four digits:	23 06	Locating a point within a 1,000-m square.
Six digits:	233 065	Locating a point within a 100-m square (football field size).
Eight digits:	2337 0651	Locating a point within a 10-m square (modest size home).
Ten digits:	23371 06519	Locating a point within a 1-m square (man hole size).

A modest size home can be found or identified in a local area with only an 8-digit grid.

Complete USNG value:	18S UJ 2337 0651	- Globally unique.
Without Grid Zone Designation (GZD):	UJ 2337 0651	- Regional areas.
Without GZD and 100,000-m Square ID:	2337 0651	- Local areas.

This illustrates how nationally consistent USNG coordinates are optimized for local applications. They serve as a universal map index value in a phone or incident directory for field operation locations. Unlike classic atlas grids (i.e. B3), these can be used with any paper map or atlas depicting the national grid and in web map portals such as the Washington, DC GIS (http://dcgis.dc.gov).

They can also be used in consumer GPS receivers to directly guide you to the location. This is especially beneficial at night, in heavy traffic, or major disasters when street signs are missing.

Point of Interest	Street Address	USNG Grid: 18S UJ	Telephone: (202)
Subway Sandwich & Salads	2030 M St., NW	2256 0826	223-2587
Subway Sandwich & Salads	430 8th St., SE	2698 0567	547-8200
Subway Sandwich & Salads	3504 12th St., NE	2740 1120	526-5999
Subway Sandwich & Salads	1500 Benning Rd, NE	2815 0757	388-0421

Ed: 20080420-USNGInstruct No2, page 2 of 2

US National Grid (USNG) Coordinates: *World wide context.*

Information Sheet 2/1 in this series. FGDC-STD-011-2001 From www.fgdc.gov/usng

The example below locates the Jefferson Pier at USNG: 18S UJ 23371 06519.

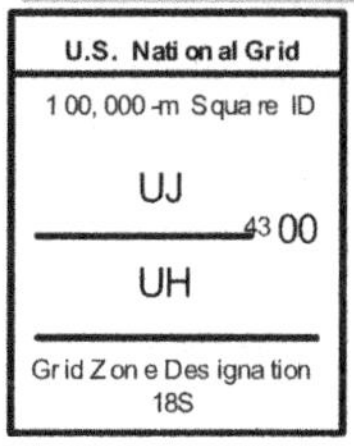

A USNG value has three components.

Some maps may give this leading information in a grid reference box.

Grid Zone Designation (GZD):
6° x 8° longitude zone / latitude band.
100,000-m Square Identification:

Grid Coordinates:
Read ***right***, then ***up***

"Read right, then up."

USNG values have three components as seen above. The Grid Zone Designation gives a USNG value world-wide context with 60 longitudinal zones each 6° wide. Zones 10 - 19 cover the conterminous U.S. as seen below left. UTM zones are divided into 8° latitudinal bands. Together these 6° zones and 8° bands compose Grid Zone Designations. Example: 18S

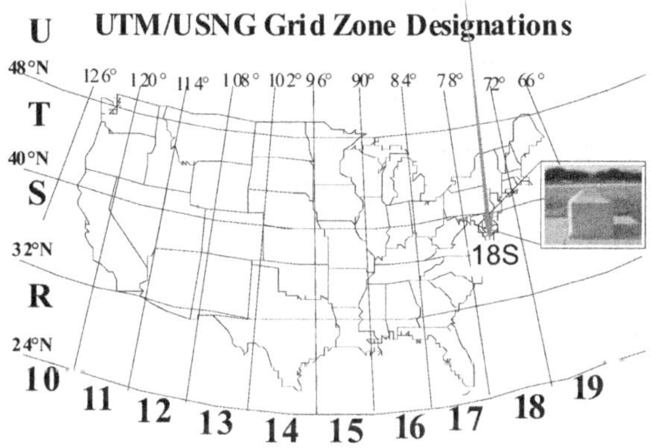

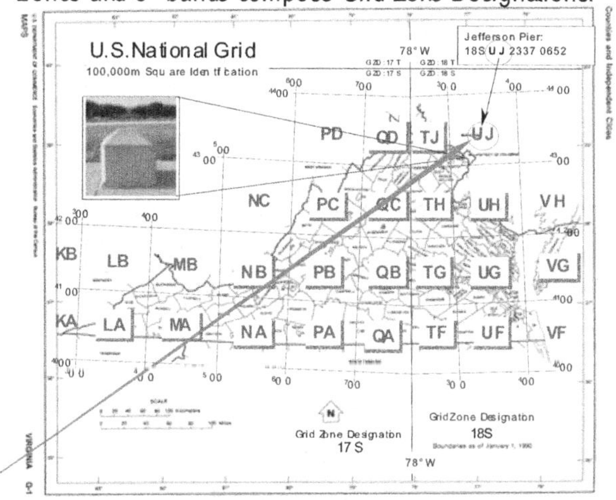

100,000-m Square Identifications
Example: UJ

GZDs are further subdivided into 100-km x 100-km squares with 100,000-m Square Identifications. In this example, the Jefferson Pier is located in UJ. These squares are organized and lettered so they do not repeat themselves but every 18°, which is approximately 1,000 miles in the mid-latitudes. The illustration at right depicts how far one must go before the letters UJ repeat. In the conterminous U.S. this ensures a given value such as UJ 2337 0651 is unique out of the entire state it is located in – as well as all surrounding states.

The Power of Truncated USNG Values

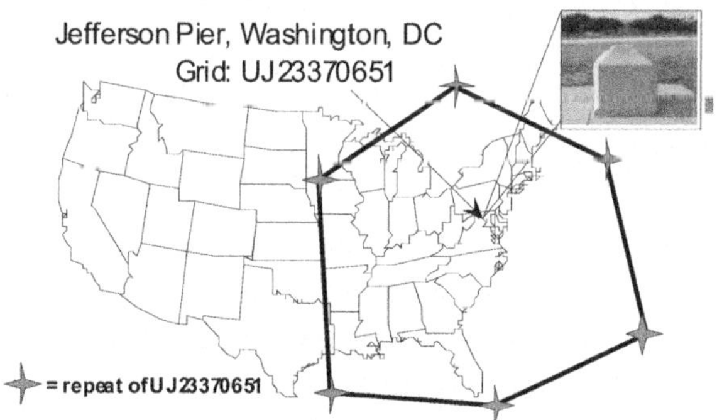

Each 2 letter/8 digit USNG value (10-m posting) in the outlined area is unique.

In general, people in a local community may use the grid coordinates alone – for example: 233 065. The same numbers recurs about every 60 miles but normally that will not cause a problem when the general location is understood. This is similar to the way you tell someone only the last digits of a phone number when the area code is obvious. If there is a possibility of confusion include the letter pair also – for example: UJ 233 065. A letter pair recurs about every 1000 miles so even in a disaster relief effort there should be no other point with those coordinates nearby. A complete USNG reference such as 18S UJ 233 065 is nationally and globally unique. Typically a GPS receiver or other electronic device requires a complete USNG reference since unlike a human it does not intuitively understand the general location from context. You should always give a complete USNG reference whenever abbreviated coordinates might not be clear or when listing them on letterhead, a business card or advertisement.

Latitude-Longitude

> **Standard Latitude/Longitude format for CISAR operations**
>
> The standard Latitude/Longitude format for CISAR operations is Degrees, Minutes, Decimal Minutes (DD° MM.mm').
>
> Latitude is always read and written first noting "North" since the U.S. is north of the Equator. Longitude is always read and written last noting "West" since the U.S. is west of the Prime Meridian.
>
> When speaking Latitude and Longitude coordinates for 39° 36.06'N by 76° 51.42'W. Latitude and longitude is stated as:
>
> *Three nine degrees, three six decimal ero six minutes North by seven six degrees, five one decimal four two minutes est.*
>
> The words, "degrees," "minutes," and "decimal" must to be spoken.

Latitude and longitude is used by aircraft and boats during CISAR operations. The Latitude-Longitude is a geographic coordinate system used for locating positions on the Earth's surface. Latitude and longitude are an angular measurement in degrees (using the symbol, " ° "), minutes (using the apostrophe symbol, " ' "), and seconds (using the quotation symbol, " " ").

Lines of latitude are horizontal lines shown running east-to-west on maps and are known as "Parallels," due to being parallel to the equator. Latitude is measured north and south ranging from 0° at the Equator to 90° at the poles (90° N for the North Pole and 90° S for the South Pole).

Lines of longitude are vertical lines shown running north and south on maps and are known as "Meridians," intersecting at the poles. Longitude is measured east and west ranging from 0° at the prime meridian to +180° East and -180° West.

Latitude and longitude can be read and written in three different formats:

- Degrees, Minutes, Decimal Minutes (DD° MM.mm');
- Degrees, Decimal Degrees (DD.DDDD°); and
- Degrees, Minutes, Seconds (DD° MM' SS").

Global Area Reference System

The Global Area Reference System (GARS) is a standardized geospatial **area** reference system for military and civil SAR application, and is based on lines of longitude and latitude. GARS provides a common language between the components and simplify communications.

GARS is a replacement for the sectional gridding so as to provide a current system that covers the entire United States with no overlapping names or coordinates. It is used to provide a grid-type search when conducting spot searches apart from the Air Search Area Definition (ASAD) search.

How GARS works.

- GARS is a worldwide system that divides the earth's surface into 30-minute by 30-minute cells.
- Each cell is identified by a five-character designation. (ex. 006AG).
- The first three characters designate a 30-minute wide longitudinal band. Beginning with the 180-degree meridian and proceeding eastward, the bands are numbered from 001 to 720, so that 180 E to 179 30'W is band 001; 179 30'W to 179 00'W is band 002; and so on.
- The fourth and fifth characters designate a 30-minute wide latitudinal band. Beginning at the south pole and proceeding northward, the bands are lettered from AA to QZ (omitting I and

O) so that 90 00'S to 89 30'S is band AA; 89 30'S to 89 00'S is band AB; and so on.

- Each 30-minute cell is divided into four 15-minute by 15-minute quadrants. The quadrants are numbered sequentially, from west to east, starting with the northernmost band. Specifically, the northwest quadrant is "1"; the northeast quadrant is "2"; the southwest quadrant is "3"; the southeast quadrant is "4".
- Each quadrant is identified by a six-character designation. (ex. 006AG3) The first five characters comprise the 30-minute cell designation. The sixth character is the quadrant number.
- Each 15-minute quadrant is divided into nine 5-minute by 5-minute areas. The areas are numbered sequentially, from west to east, starting with the northernmost band.
- The graphical representation of a 15-minute quadrant with numbered 5-minute by 5-minute areas resembles a telephone keypad.

Each 5-minute by 5-minute area or keypad "key" is identified by a seven-character designation. The first six characters comprise the 15-minute quadrant designation. The seventh character is the keypad "key" number. (ex.006AG39).

Figure 4-7-3 below graphically depicts GARS, and Figure 4-7-4 is an example GARS overlay.

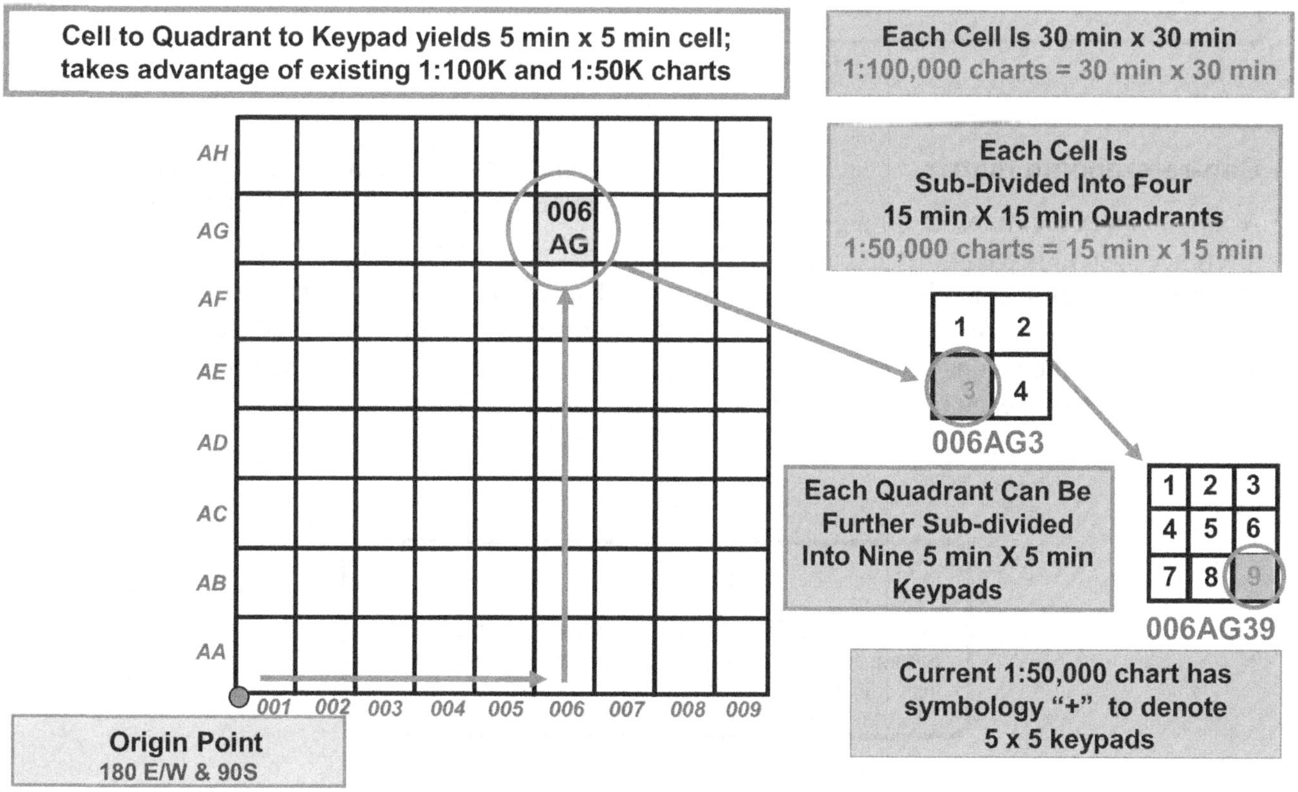

Figure 4-7-3: Global Area Reference System (GARS)

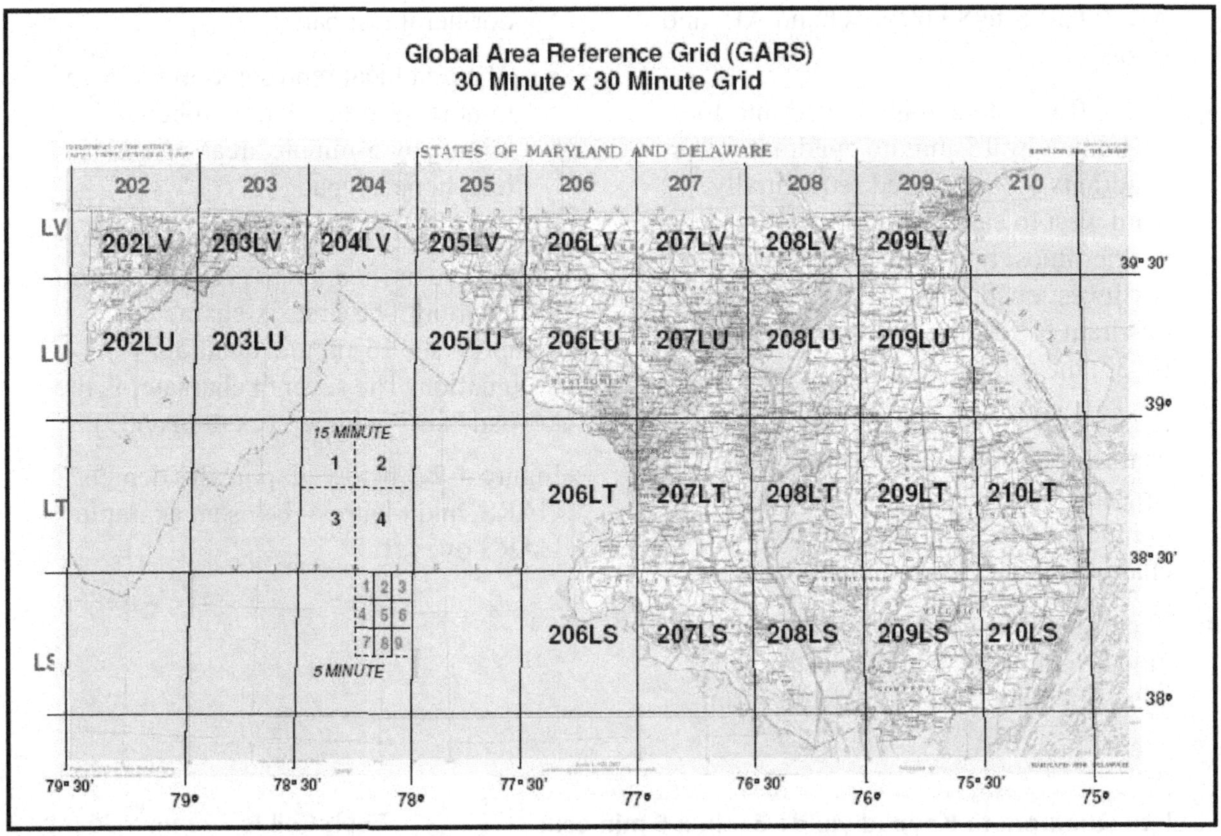

Figure 4-7-4: GARS Overlay

Geo-referencing matrix

A fundamental requirement for a geo-reference system is the ability to easily interface between the incident command, the land SAR responder (or maritime SAR responder) and the aeronautical SAR responder. Because each has unique geo-referencing requirements, effective interface between each component is vital to a successful SAR response.

The geo-referencing matrix minimizes confusion and provides guidance on what geo-referencing system each SAR responder should be using.

Map Datum

North American Datum 1983 (NAD 83) and World Geodetic System 1984 (WGS 84) are equivalent at scales smaller than 1:5000.

Table 4-7-1: National SAR Committee's Geo-referencing Matrix

Georeference System User	United States National Grid (USNG)	Latitude/Longitude DD-MM.mm[1]	GARS[2]
Land SAR Responder[3]	Primary	Secondary	N/A
Aeronautical SAR Responders[4]	Secondary	Primary	As Needed[10]
Air Space Deconfliction[5]	N/A	Primary	N/A
Land SAR Responder/ Aeronautical SAR Responder Interface.[6]	Primary	Secondary	N/A
Incident Command:			
Air SAR Coordination	Secondary	Primary	N/A
Land SAR Coordination	Primary	Secondary	N/A
Area organization and accountability[7]	Secondary	Tertiary	Primary

[1] During SAR operations (and to avoid confusion) Latitude and Longitude should be in one standard format: DD-MM.mm. If required, use up to 2 digits to the right of the decimal. If required, allow 3 digits in the degrees field for longitude (i.e., DDD-MM.mm). Do not use leading zeros to the left of the decimal for degrees or minutes that require fewer than the maximum number of possible digits to express their value. The minimum number of digits is always one, even if it is a zero. (Example: Recommended: 9-0.3N 4-2.45W; Not Recommended: 09-00.300N 004-02.45W).

[2] GARS: Global Area Reference System.

[3] Land SAR responders use U.S. National Grid; however, a good familiarity with latitude and longitude is necessary to ensure effective interface between Land and Aeronautical SAR responders (Note: Land SAR includes SAR on flooded terrain).

[4] Aeronautical SAR responders will use latitude and longitude for SAR response. However, aeronautical SAR responders that work directly with Land SAR responders should understand the U.S. National Grid system for effective Land SAR/Aeronautical SAR interface.

[5] Air space deconfliction will *only* be implemented and managed using latitude and longitude.

[6] Aeronautical SAR responders working with Land SAR responders have the primary responsibility of coordinating SAR using USNG. However both groups must become familiar with both georeference systems.

[7] Describes the requirement for providing situational awareness of SAR operations geographically to Federal, military, state, local and tribal leadership. Provides for quick reference to send SAR resources closest to incident.

This page intentionally left blank.

Section 4-8: SAR Case Documentation

ICS Forms

SAR Action Plans

Reports

Records Management

ICS Forms

The NIMS/ICS forms are designed to assist emergency response personnel in the use of ICS and corresponding documentation during incident operations. These forms are meant to complement existing incident management programs and do not replace relevant emergency operations plans, laws, and ordinances. They are tools for the creation of Incident Action Plans (IAPs), incident management activities, and for support and documentation of ICS activities.

An understanding of NIMS, including ICS, through training and experience, can improve effective use of these forms.

The forms accommodate essential ICS data elements and promote consistent management and documentation of incidents and facilitate effective use of mutual aid. Additional pages can be added to a form as required.

SAR Action Plans

Every search should have an action plan and/or IAP. For small searches the plan may be unwritten but for larger searches a written plan may prove essential.

A SAR Action Plan should:

- Define operational plans;
- Provide written objectives reflecting policy and needs of all jurisdictions;
- Incorporate subject profile data and missing/lost person behavior;
- Provide an organizational chart and assignment lists;
- Establish search maps delineating assignment areas;
- List operational objectives for defined periods;
- Give resource status and availability;
- Provide regular situation/status reports;
- Outline medical, rescue and recovery, communication and transportation plans; and
- Identify safety considerations.

The plan should be developed initially by the Planning Section with guidance from the SMC/IC and then briefed. The SMC/IC determines:

- What information is required;
- Reporting schedules for all the organizational elements; and
- General objectives and alternatives that define legal, policy, resource, and fiscal constraints in accordance with policies and pre-plans of the involved jurisdictions.

Reports

An important component of a SAR operation is complete documentation with accurate data. It is a sound practice to prepare proper reports when an incident is completed. The SAR Coordinator/ Responsible Authority can use reports to evaluate performance, design preventive SAR plans, develop statistical models, assist with investigations, seek emergency funding, assist in tort claims, and document agency activity.

Reports should be treated with confidentiality if containing critical or sensitive information.

The following types of reports are recommended:

- *Search and Rescue Report* (commonly referred to as the "Mission Report") - Complete this report to the extent possible and file with the Case Incident Report. General (non-Privacy Act) information, such as subject behavior or statistical data may later be gleaned from this report.
- *Case Incident Report* - Complete this report with incident details and actions taken by responders. Attach relevant documents, such as Supplementary Case Incident Reports, diagrams, maps, photographs or ICS forms.
- *Aircraft Use Documents* - If the SAR incident involved use of Government-owned or chartered aircraft, complete use documents and submit through normal channels with copies filed with the Case Incident Report.
- *Training and Re-certification Reports* - Either through the SAR Coordinator/ Responsible Authority or other designated means, each program is responsible for ensuring that SAR training provided by the organization or otherwise obtained by employees is documented. These documents are the administrative record for each SAR provider's re-certification, program audits, statistical information, and other related purposes.

Records Management

Managers should maintain records in cases where emergency service is provided by SAR personnel to protect the subject and can be invaluable should legal questions arise. Each agency has its own criteria and regulatory guidance for records management and disposition. Personal information about the subject should be protected according to applicable law, including, in the case of Federal Agencies, the Privacy Act.

Section 4-9: Briefing and Debriefing SAR Teams

Briefing

Debriefing

Briefing

Briefings give the direction, instructions, and orientation to performing any action in the search effort. Briefings take place throughout the SAR process and are important to convey appropriate information in a timely manner. All involved personnel must be briefed before commencing SAR operations.

Functional areas that briefings should cover include:

- Search situation status to date;
- Search object profile that details information necessary for those involved in the search;
- Objectives;
- Strategy;
- Tactical assignments;
- Safety;
- Media, family, friends, or medical authority protocols;
- Clothing, possessions, or other incidentals known to be with the search object;
- Expected terrain, vegetation, and weather related visual impacts;
- Examples of visual cues and clues to look for;
- Search techniques; and
- Debriefing instructions with information guidelines, check in/check out procedures, etc.

Briefings are usually conducted as per the following:

- The SMC/IC briefs the responsible SAR authority;
- The outgoing SMC/IC at shift change briefs the incoming SMC/IC;
- The Operations lead will brief and debrief team leaders for all assignments;
- The media should be briefed at designated times by either the SMC/IC or the Public Information Officer (IO);
- The SMC/IC or family liaison officer should brief family members on a regular, scheduled basis; and
- Field team leaders brief team members before each assignment.

Briefings should be kept short, factual, and to the point.

Debriefing

SAR personnel debriefs are used to review and verify information after completion of a SAR assignment and at the conclusion of the SAR incident. The process is used to preserve the observations and impressions of the field units for documentation and the development of future plans.

Ideally, debriefs should occur as soon as possible after the search. In particular, debriefing should take place as each searcher accomplishes their respective search for that period. This is especially important in verifying that the IAP/SAR plan was accomplished as directed and expected. Any changes that deviated from the expected plan should be debriefed and may affect any future plans.

An effective SAR team debriefing should be:

- Timely;
- Written;
- Include recommendations;
- Focus on individual beliefs;
- Thorough and exact; and
- Detailed.

The debriefing should be in writing, or recorded. When written, it reduces misinformation, misinterpretation, and confusion. Also, when written it becomes part of the incident record.

Table 4-9-1 is a sample Debriefing Checklist.

Table 4-9-1: Debriefing Checklist	
Time task completed:	Name of debriefer:
Date/Time prepared:	Total time actually searching:
Time search started:	Time search stopped:
How many searchers on team:	How far apart was spacing (feet):
Size of search area searched: ____________ (acres) ____________ (square km)	
Team Performance	
Visibility/weather conditions:	
Terrain conditions:	Hazards noted:
Was task completed as assigned? YES NO (circle one) If NO, explain:	
Average maximum detection range (AMDR) (feet):	Total track line length (TLL) (feet):
Qualitative description for search: POOR AVERAGE GREAT (circle one). Narrative:	
Estimate of forward speed of search: FAST, NORMAL, SLOW (circle one):	
Were there any gaps in the search area? YES NO (circle one). If YES, explain:	
Was the team adequately equipped? YES NO (circle one). If NO, explain:	
Rate the team's composition: POOR AVERAGE GREAT (circle one). Narrative:	
Rate the team's performance: POOR AVERAGE GREAT (circle one). Narrative:	
Rate the team's morale: POOR AVERAGE GREAT (circle one). Narrative:	
Dog Tasks	
Wind direction, speed and variability:	Direction/strength of alerts:
Was the scent article adequate? YES NO (circle one). If NO, explain:	
Team Leader	
Comments/recommendations:	
Searchers observations:	
Recommendation for re-search of area:	
Additional comments (as required, continue on the back of this form)	

This page intentionally left blank.

Part 5: Search Planning

This page intentionally left blank.

Section 5-1: Search Planning Basics

Search Planning

Searching for a person in distress is an emergency operation. A basic principle of SAR is that the saving of human life is the priority and should not be hindered by jurisdictional boundaries. The sooner persons in distress are found, the better the chances they will be found alive. Over time, environmental conditions or injuries can reduce the distressed person's chances for survival significantly.

SAR responders should minimize the time required to begin a search, and then optimally allocate available search resources to find the subject in the least amount of time. This requires a balance of practical considerations and valid planning processes. Search planning requires knowledge, training, and sound judgment, along with necessary tools to quickly produce the best possible search action plan.

Careful use of proper search planning tools, information and concepts (e.g., 406 MHz distress beacon registration, radar and cell phone forensics, and optimal effort allocation techniques) substantially improves the chances that the subject can be located and rescued in time.

Search planning includes the following seven steps:

7 Steps of Search Planning

1) Evaluate situation and all available information.
2) Estimate the distress incident location.
3) Estimate the distress person's movements.
4) Using the estimated results from 2) and 3), determine the most probable location of distress person/aircraft.
5) Determine the best way to use the available search resources so the chances of finding the distress person(s)/aircraft are maximized.
6) Define search sub-areas and search patterns for assignment to specific search resources.
7) Provide a search action plan that includes a current description of the situation, search object description(s), assigns specific search responsibilities to search resources, on-scene coordination instructions, and search facility reporting requirements.

IAMSAR Manual, Volume II (Chapter 4)

When one or more searches have been conducted, information gained from these searches must be applied to planning any additional searches.

The chances of survival for lost or missing persons tends to decline rapidly as time passes, especially if they are injured or ill-prepared for hazardous weather or other conditions encountered. There are often high levels of emotional stress associated with search planning and conducting search operations.

Search Theory

Search theory is the study of mathematical methods and algorithms for developing optimal search plans. It is a branch of the broader applied science known as operations research.

Search theory focuses on:

- where to search; and
- how to search.

The goal of search planning is to deploy available SAR resources in a manner that maximizes the chances of finding the subject in the least amount of time. Finding subjects quickly saves more lives, reduces search costs, and minimizes risk to searchers. For these reasons, it is prudent to use the planning methods, tactics, and strategies that both science and experience have to offer.

Search Objectives

Search objectives lead to strategies, and strategies to tactics.

Objectives should be specific, measurable, achievable, realistic, and time-referenced. For example, "confinement" is a strategy search planners can use to keep the subject from leaving the search area. Once search planners determine their search strategy, a search tactic can be selected. For example, track traps and perimeter searches are possible tactics for a "confinement" strategy. Figure 5-1-1 below provides a summary of search objectives, strategies, and tactics.

Figure 5-1-1: Overview - Search Objectives, Strategies, and Tactics.

The search tactics chosen must support the objectives. The search tactics must be matched with the skills and characteristics of the search resources.

Developing objectives is vital in defining the search strategy and in turn selecting the right resources and utilizing them in the most effective manner.

Search Tactics: General Considerations

Regardless of how the search resources are utilized, search tactics should abide by the basic guidelines for search patterns found in Table 5 -1-1 below to build confidence in the search results, or to allow a quick transition into the next operational period.

(Note: Refer to Section 5-4 (Resource Strategies and Tactics) for information on specific search tactics and patterns for various search resources.)

Table 5-1-1: Search Pattern General Considerations	
Ease of performance	• Easy to navigate; • Simple for searchers to find and stay within their assigned boundaries; and • Simple for searchers to stay properly spaced in order to cover their area consistently.
Safe	• SAR Responder safety is paramount.
Ease of change	• Easy to reassign resources based any updates to the search object's position or other clues; • Easy to re-search, especially in a way that lets searchers see clues from a different perspective (a different spacing, different direction, etc.); and • Easy to expand the search area as required.
Ease of evaluation	• Resources cover search area in a method that allows for the determination of how well it was searched and to facilitate follow-on decisions.

Core Search Planning Considerations

There are core search planning considerations that must be weighed by search planners for every mission (see Table 5-1-2).

Each SAR mission is unique, with unique circumstances and challenges. Table 5-1-2 should only be the starting point for determining planning considerations for a specific SAR mission.

Baltimore County, Maryland. SAR Briefing. (Photo: Cole Brown)

Table 5-1-2: Core Search Planning Considerations	
Who, or what, is being searched for?	**Who** or **what** is being searched for will impact the decision on how to search. Consider the differences among: • Mobile vs. immobile person; • Responsive vs. unresponsive person; • Lost child vs. lost adult; and • Missing aircraft vs. distress beacons.
What SAR resources are available to conduct the search?	**What** SAR resources you plan to search with will impact the decision on how to search. Consider the differences among: • Trained SAR responders vs. untrained volunteers; • Helicopters vs. light fixed wing aircraft; • Aircraft equipped with a single observer vs. aircraft with multiple scanners; and • Day vs. night search capabilities.
Where is the search being conducted?	**Where** the search is conducted will impact the decision on how to search. Consider the differences among: • Meadows vs. forests; • Flat areas vs. hills or mountains; and • Campgrounds and parks vs. wilderness areas.
When is the search being conducted?	**When** the search is conducted will impact the decision on how to search. Consider the following: • Time of day affects sun angle which affects visibility of clues; • Time of day affects searcher freshness and thoroughness; • Time of year affects sun angle, temperature, and length of day; and • Time since distress or LKP/PLS affects the freshness, visibility, and even the existence of clues.

Other Questions That Should Be Asked

SAR managers and planners should consider the following additional questions when planning the search:

- Are the appropriate SAR resources available quickly?
- Will nightfall/weather change the search method and the resources used?
- How will the balance be maintained between the desire to "get out there and do something" and the potential destruction of clues by Rapid (hasty) Search teams?
- How will the balance be maintained between the desire to find the person(s) in distress right away and the desire to limit their travel in the event they are not found right away?

Preplanning

Preplanning represents guidelines and current information set up before a search to allow a quick and effective response, strategic decisions to be made outside the heat of the moment, and flexibility within a well thought out selection of reasonable actions. Preplanning is an important method to help ensure the initial SAR response is quick and appropriate.

Decisions specific to a particular search, including tactical operations, should be in the search or incident action plan. The processes to be followed for building the action plan should be in the preplan.

The SAR preplan must be kept simple and up-to-date and should be reviewed and updated periodically (e.g., before and after exercises and after an actual SAR incident). The planning process needs periodic check-ups and reviews for efficacy and accuracy of information after every use.

Six major elements to be considered when drafting any SAR preplan include:

- Environment – maps, risks, hazards, attractions, past incidents, etc.;
- Resources – who/what is available, capabilities, limitations, contacts, etc.;
- Initial Notification – guidelines, questions for initial call, checklists, who is in charge, etc.;
- Action Plan Development – guides and procedures, SAR organization, planning search strategy and possible tactics, managing search assets, etc.;
- Support Requirements – logistics, communications, medical, etc.; and
- Post-Mission Actions – demobilization, lessons learned, reporting requirements, etc.

(Note: For a more complete example of SAR Pre-plan elements, see Appendix C.)

Needs Assessment

Each agency that is party to a SAR preplan should perform a needs assessment. The needs assessment is used to identify current and projected SAR mission requirements to comply with preplans, rather than to justify current operating conditions. The process is a tool used to assess the current condition of the SAR program and then determine whether the provided services are consistent with contemporary standards. A needs assessment identifies and evaluates:

- Available internal and external resources;
- The agency's SAR workload;
- Requirements for training and certification;
- Rescue capabilities and response times;
- Location and capability of local area resources;
- Fiscal resources;
- SAR communications; and
- Special considerations (e.g., mutual aid, out of area response, geographic location, etc.).

Risk Assessment

A risk assessment is performed to avoid jeopardizing the success of the mission by unnecessarily endangering the lives of both responder(s) and the subject/survivor.

(Note: See the CISAR Addendum for additional information concerning risk assessments.)

This page intentionally left blank.

Section 5-2: Search Planning Methods and Tools

Elements of Planning

Key Concepts

- *Search Area*
- *Regions*
- *Segments*
- *Probability of Detection (POD)*
- *Sweep Width (W)*
- *Effort (z)*
- *Parallel Sweep (or Route) Search*
- *Coverage (C)*

Probability of Detection (POD) – Continued

- *Cumulative POD (POD_{cum})*

Probability of Success (POS)

Overall Probability of Success (OPOS)

- *Cumulative Overall Probability of Success ($OPOS_{cum}$)*

Probable Success Rate (PSR)

Elements of Planning

The goal of search planning is to deploy available search resources in the most effective way to maximize the chances for locating the search subject in the shortest amount of time. A search action plan should be developed that will make optimal use of available resources despite the uncertainties that necessarily accompany every search.

A search plan should include the following:

- Location and dimensions of the search area and the search sub-areas;
- Optimal allocation of available effort (e.g. searcher-hours) to the appropriate sub-areas; and
- Assignment of specific SAR resources to their corresponding sub-areas.

Key Concepts

Search planners should be familiar with the definitions introduced in the following subsections:

- Search area;
- Regions;
- Probability of Containment (POC);
- Segments;
- Probability of Detection (POD);
- Sweep Width (W);
- Effort (z);
- Parallel Sweep (or Route) Search; and
- Coverage (C);

> **Key Concept:**
>
> In land SAR planning, the tracking of the values for Probability of Containment (POC), Probability of Detection (POD), and Probability of Success (POS) as presented in this Addendum are related to Area/Grid searching.

Search Area

The "search area" is the physical area the search planner believes is likely to contain the search object. Search planning is a process of elimination. The search planner must attempt to eliminate as much area as possible without discarding locations where the search object may be. In developing the search area, search planners should consider the search object's:

- Departure point;
- Intended destination;
- Possible or likely routes;
- Historical data concerning the area;
- Personal habits and inclinations of the missing person;
- Lost person behavior;
- Location where the object was last seen, reported, or otherwise known to be; and
- Location where the objective was last seen.

Search planning largely depends on investigation. Investigative efforts should continue unabated throughout the search until the object is located. Investigation provides a means to gather and clarify information, as well as deal with clues or information which was not previously included in determining the search area.

Regions

A region is a defined subset of the search area within which there is a nearly uniform distribution of the likelihood of containing the search object. In other words, the search object is equally likely to be at any point in a given region. Regions and region boundaries are based only on the factors that affect the search object, and the change in likelihood of the search object's location on one side of the region boundary versus the other side. A region boundary means the search planner believes there is a reason why the search object is more likely to be on one side of the boundary than the other.

Probability of Containment (POC)

> **Probability of Area (POA) and Probability of Containment (POC)**
>
> POA is a term used mainly by land SAR practitioners today. The terms POA and POC are synonymous, with POC being the preferred term with international recognition.
>
> POC will be used in this Addendum for SAR planning.

Regions of probability and their Probability of Containment (POC) values are determined by carefully considering all available information, adding reasonable assumptions to fill the information gaps and create plausible scenarios. Each scenario is then carefully examined in turn to see where it leads in terms of more probable and less probable search object locations.

Searching a large area can be labor intensive with limited search resources available. Rarely will the general search area be small enough for the available resources to search the entire area rapidly and with a high probability of finding the search object in a short time.

Probability of Containment (POC) for any portion of the search area is defined as the probability of the search object being in that portion of the search area. POC is only one of the factors to consider when deciding where resources should be deployed.

Segments

In contrast to regions of probability, segments are sub-divisions of the search area based on the logistical and operational issues associated with conducting the search itself. In other words, segments are based on the searchers and their capabilities, not the search object. Segments have associated POC values by virtue of being a portion of the general search area, but their boundaries and sizes are not based on concerns about where the search object may be located. For example, a segment should be small enough that it can be searched by the intended type of SAR resource with a reasonable coverage in a specified amount of time.

For aircraft, segments are generally much larger than those assigned to ground teams and are usually limited by the aircraft's on scene endurance.

Probability of Detection (POD)

A successful search is based on searching in the right location and detecting the search object. Probability of Detection (POD) is defined as the probability of finding the search object, assuming it is in the segment being searched. POD is determined at the segment level and depends on two factors:

1. How easy or difficult it will be for the object to be detected by searchers in the segment; and
2. How much search effort is expended in the segment being searched in relation to the segment's size.

Sweep Width (W)

The ease or difficulty of detecting the search object is measured by a quantity called Effective Sweep Width (ESW).

This quantity may also be referred to as Sweep Width (W), and is most easily understood by considering ESW or W as an "index of detectability." A measurement of how easy or hard it will be to detect a given search object in a specific environment with a particular sensor. Sweep width is affected by the following factors.

- Nature of the search object;
- Terrain, vegetation, visibility, and weather conditions of the search area;
- Altitude of the search aircraft above ground level;
- Search speed (v);
- Searcher fatigue;
- Searcher capabilities; and
- Sensors being used in the search (e.g., unaided eye, night vision goggles, infrared, air scent dogs, etc.).

Conceptually, sweep width is the distance (equally distributed to the left and right of the sensor) at which the likelihood the sensor will detect an object beyond that distance is equal to the likelihood the sensor will fail to detect an object within that distance. See Figure 5-2-1 below.

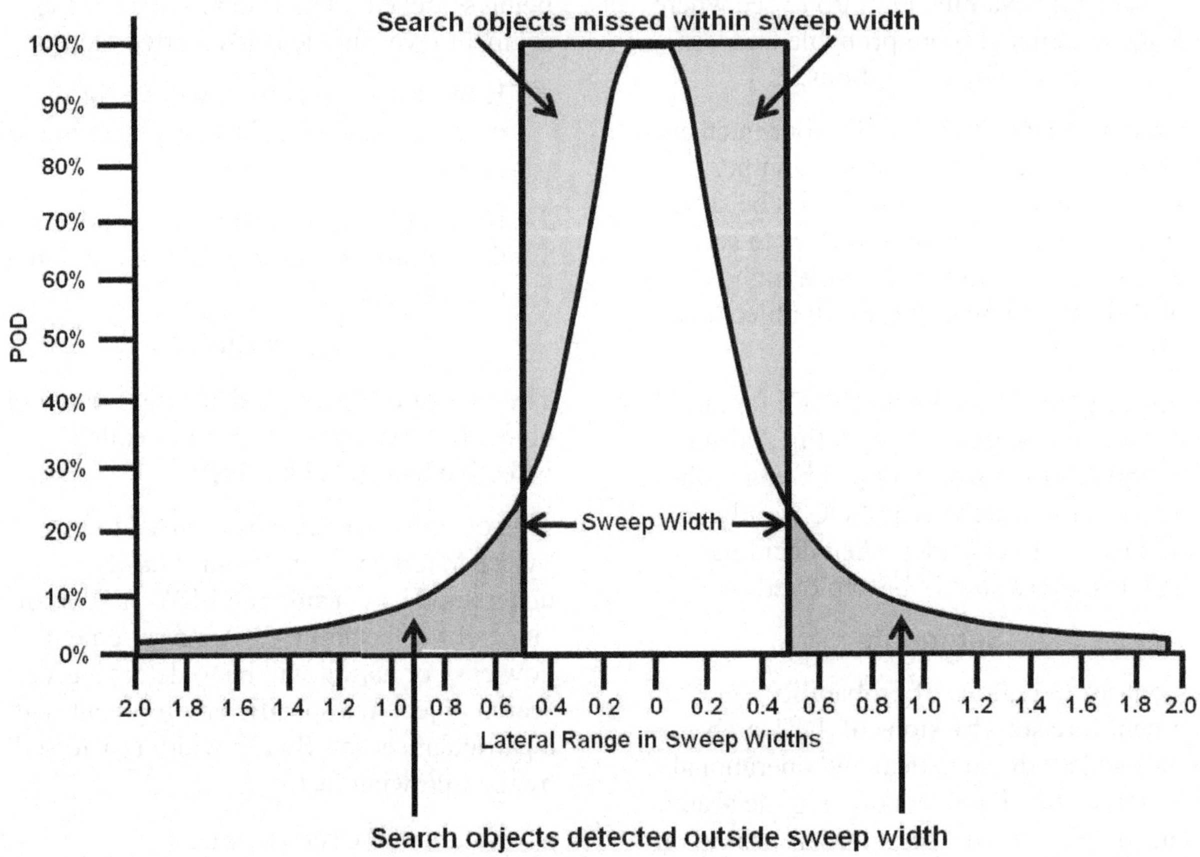

Figure 5-2-1: Sweep Width.

Sweep width recognizes the likelihood of not detecting objects near the observer, along with the likelihood of detecting objects at a longer range from the observer. Sweep width does not indicate the maximum distance at which an object might be sighted, nor that an object would be sighted within a specific range.

Sweep width can only be determined through extensive field experimentation during which the numbers of detections and misses are counted for a large number of sensors (searchers) at a wide variety of distances from objects. A lateral range curve is generated from the detection data and is used to find the sweep width for the tested sensors as used under the conditions in the experiment. The lateral range curve graphs the probability of detecting an object against the distance from the searcher to the object at the closest point of approach.

Sweep width is very specific and dependent upon the conditions experienced during the search. For instance, visibility, type of object, vegetation, and terrain may inhibit detecting the object. When a sensor is tested in a variety of environments, tables and corrections can be constructed in order to make sweep width more widely applicable.

While the search planner may be able to make some choices regarding which searchers and sensors are assigned to a search segment, the planner cannot control the nature of the search object, the search conditions in the area, or how detectable the search object will be at the time and place of the search. In short, the search planner cannot control or adjust the effective sweep width.

Note: Further information on conducting ground sweep width studies can be found on the National SAR Committee website (www.uscg.mil/nsarc).

The search planner does decide how much of the available effort should be allocated to each of the various segments of the search area. This is the essence of the search planner's job.

Effort (z)

Effort (z) is the total distance searchers travel while searching in their assigned segment. It is often useful to compute or estimate effort as the total number of searcher-hours spent searching (X) the average search speed. The relationship among speed, distance, and time may also allow searcher-hours to be used as a measure of effort.

Effort is also denoted by the abbreviation "*TTL*", which stands for total [searcher or aircraft] trackline length.

Effort is computed as number of searcher-hours spent searching times the average search speed:

$$TTL = (t) \times (n) \times (v)$$

Where t is time spent searching, n is number of searchers, and v is speed while searching.

For example, the "Effort" for 20 searchers searching a segment for 5 hours at an average speed of .5 mph is computed as:

TTL (Effort): 5 hours X 20 searchers X .5 mph = 50 miles

Coverage (C)

Since both W and TTL (or z) have units of length, their product has units of area. We may compute a value called the *area effectively swept (Z)*:

$$\text{Area Effectively Swept} = W \times TTL$$

The ratio of "area effectively swept" to the physical area that was searched is called *coverage* or *coverage factor (C).*

Coverage may be thought of as a measure of "thoroughness" or "effort density," how much searching effort is expended per unit of area in a given locale. The key concept being: the higher the coverage, the more thorough the search. Coverage is a ratio of like quantities and has no units.

Coverage factor is the result of comparing how a sensor should perform generically with how it is used on a specific search.

The formula for Coverage is as follows:

$$\text{Coverage (C)} = \frac{\text{Total Effort (Z)}}{\text{Area Searched}} = \frac{\text{Sweep Width X Speed X Time}}{\text{Area Searched}} = \frac{Wvt}{\text{Area}}$$

(Note: Time (t) is total searcher hours for the ground environment or hours spent searching for the air environment.)

Parallel Sweep (or Route) Search

Parallel Sweep Searches (also known as Route Searches in the ground environment) is defined as a systematic search in which search team members follow tracks parallel to a side boundary and maintain a predetermined separation (see Section 4-4).

Parallel Sweep Search Coverage can be computed using the following formula:

$$\text{Coverage (C)} = \frac{\text{Sweep Width}}{\text{Searcher Spacing}} = \frac{W}{S}$$

This simplified formula works best in an air search environment, when conducting a parallel sweep search, or a strict grid search in the ground environment.

Key Concepts: Summary

The important points in this discussion are the following:

- Higher coverage produces higher POD, lower coverage produces lower POD;
- There is a direct relationship between effort and coverage: Doubling the coverage of a segment requires doubling the effort assigned to that segment;
- Any increase in POD requires an increase in coverage with a corresponding increase in effort; and
- For a given level of effort, larger areas or more segments can be searched at lower coverage, smaller areas or fewer segments can be searched at higher coverage. A "mixed bag" approach can also be used where some segments receive a high coverage, some moderate coverage, some low coverage, and some segments may not be searched at all (coverage = 0).

Probability of Detection (POD) - Continued

As stated previously, POD is the probability of detecting the search object (as a function of the effort expended in that search area) given that the object is in the area searched.

POD is the statistical measure of detection performance. It is a function of:

- Sweep width (W);
- Level of effort (z); and
- Size of the area searched.

POD allows the search planner to have an estimate of how likely an SRU in a given segment is to detect the search object if it is in the search segment. It has a predictive power for search planning, and after a search period and debrief, planners can re-compute POD for conditions searchers report in the search segment.

POD is a "conditional probability." The "condition" is an assumption that the search object is definitely in the area searched. POD answers the following question:

If the search object was in the searched area at the time of the search, what was the probability of detecting it?

The relationship between Coverage (C) and POD is given by the following equation and its graph in Figure 5-2-2:

$$POD = 1 - e^{-C}$$

In this equation, "e" is the base of the natural logarithms (approximately 2.718282).

(Note: The value of "e" raised to a power can be found on most scientific calculators as the e^x or "EXP" button.)

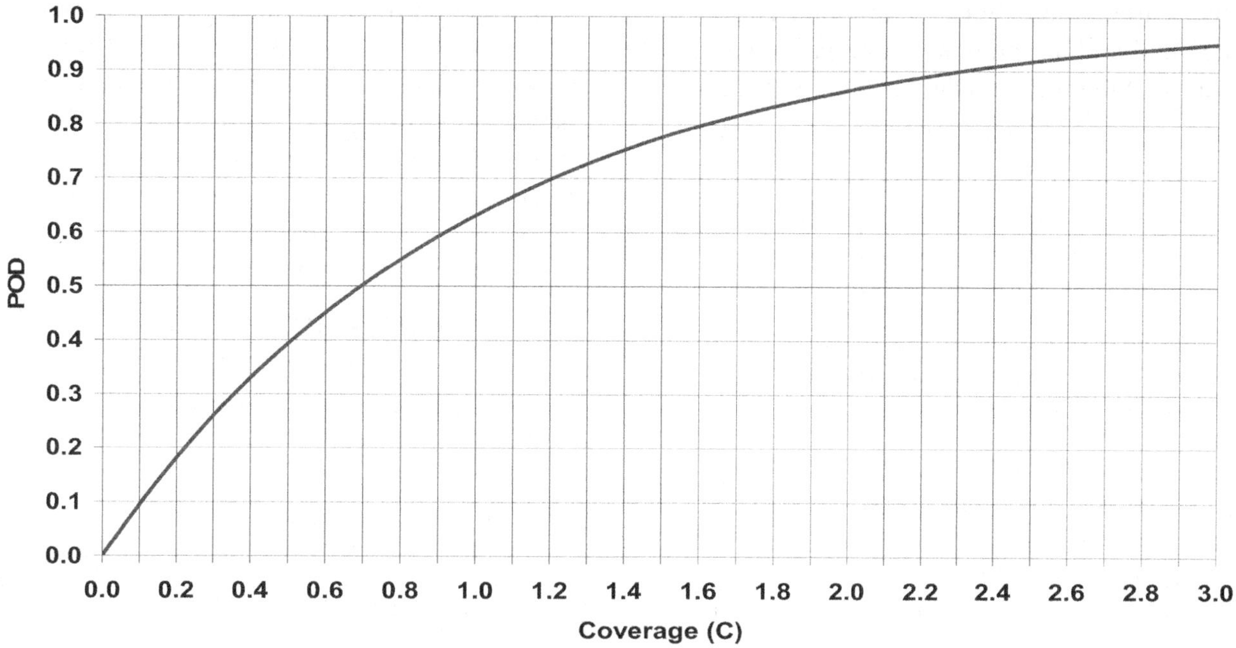

Figure 5-2-2: Chart - POD vs. Coverage (C)

The graph above depicts diminishing returns. As effort is added and coverage increases, the POD also increases, but more and more slowly as the amount of effort and the corresponding coverage reach higher and higher levels.

Cumulative POD (POD_c).

If the coverage factor for each search period in a segment is known, the graph and equation may be used to compute cumulative POD (POD_c) for two or more searches of the same segment.

For example, if an initial search is completed with coverage of 0.5 and the same segment is searched a second time with coverage of 1.0, then the "cumulative coverage" for both searches is 1.5 and the cumulative POD is about 78%. A third search with coverage of 0.8 would bring the "cumulative coverage" to 2.3 and the cumulative POD to about 90%.

Search planners need to realize cumulative coverage and POD for a particular search segment only address that segment, and do not account for whether the search object is *in* the segment searched. Planners need more information to predict/determine the likely success of a search.

Probability of Success (POS)

When planning a SAR mission, the planner should always be concerned with obtaining the *optimal* SAR plan. In a SAR context, a plan is considered optimal if it maximizes the *Probability of Success (POS),* or the probability of finding the search object.

POS is determined by two quantities:

- POD – the probability of detecting the search object in a search area; and
- POC – the probability that the search object is within the area being searched.

This can be written mathematically:

POS = POC X POD

POS is a statistically generated measure of search effectiveness and is the probability that a given search will succeed in locating the search object. POS is expressed as a decimal and has no units. For a particular search, POS answers the following question

> *If the POC, and POD values in a given scenario are an accurate reflection of the available information and data, what is/was the probability of finding the search object in a given search period?*

Cumulative *POS* (POS_c) is a measure of search effectiveness to date and answers the following question:

> *If the POC and POD values in a given scenario are an accurate reflection of the available information and data, what is the probability that the search object would have been found by now?*

Achieving a high POS_c value without locating the search object is an indication that either the search object cannot be detected (e.g., on the bottom in deep water, hidden by an obstruction, etc.) or that the POC values, and/or POD values in a given scenario are suspect. A thorough review of all the available information is required to determine whether POS_c has been interpreted, computed, and used correctly.

The following compares POS, POD and POC:

- POS measures search effectiveness;
- POD measures search sensor detection performance; and
- POC measures the likelihood the search object(s) are contained in a particular area.

POS is the most effective way of measuring search effectiveness. It is the measure used in the IAMSAR Manual, the National SAR Supplement (NSS), and in the NSS's U.S. Coast Guard Addendum.

Although POD has been in the search planning vocabulary and used with the manual search planning method for many years, POS provides a much greater measure of search quality. This is because POD only measures detection effectiveness. That is, it is used to estimate how well a search area was searched, but it does not incorporate the likelihood that the search object will actually be in the particular area searched.

POS, POD, and POC

(POS = POC X POD)

The following points need to be understood concerning the relationships between POS, POD and POC:

- Searching an area that has no chance of containing the search object (POC = 0) will not be successful no matter how high the POD. Even if POD was 100% the POS is still zero (0 x 1 = 0);
- If there is a 50% chance of the search object being in an area, then searching that area with a coverage factor of 1.0 (POD of 63%) produces a POS of 32% (0.5 x 0.63 = 0.32). Even if POD was 100%, the POS for this search rises to only 50% (0.5 x 1 = 0.5) and no further because there is still a 50% chance that the search object was not in the search area.

POS balances options of looking very carefully in a small area for the object, against looking less thoroughly over a larger area for the same object.

Overall Probability of Success (OPOS)

The POS for any given segment being searched is given by the following equation:

$$POS_{segment} = POC_{segment} \times POD_{segment}$$

However, the search planner's attention cannot be limited to a single segment. For each operational period, the search planner's job is to deploy available resources so the overall POS (OPOS) is maximized. The OPOS is simply the sum of all the individual segment POS values:

$$OPOS_{cum} = POS_{segment\ 1} + POS_{segment\ 2}$$

There is an important difference between the OPOS based on all searched segments and individual segment POD and POS values. For any single segment, the only way to increase POS is to increase POD, and the only way to increase POD is to increase the level of effort expended in the segment.

For a limited amount of total available effort, which is the usual situation, the level of effort in one segment cannot be increased without taking effort from some other segment(s). Putting all the available effort into just one segment would maximize its POS, but it is virtually certain that a higher OPOS could be obtained from allocating the available effort so that multiple (but not necessarily all) segments each receive some searching.

The manner in which the available effort is apportioned among the search segments makes a difference in the value of the OPOS for that operational period. The best chance for finding the search object with the available resources depends upon how those resources are distributed among the search segments.

For a given amount of available effort, there is an optimal effort allocation among the search segments that will produce the maximum OPOS. Search theory provides a means for:

- Computing the optimal effort allocation; and
- How much of the total available effort should be placed in each segment in order to achieve the maximum possible OPOS value.

Cumulative Overall Probability of Success ($OPOS_{cum}$)

The cumulative overall POS ($OPOS_{cum}$) is the accumulated sum of all POS values for all searching done to date. When $OPOS_{cum}$ reaches a high value but the search object remains unlocated, search planners need to re-evaluate the case information to ensure that:

- No possible scenarios were overlooked;
- All of available information was correctly interpreted;
- The SAR case assumptions were reasonable; and
- All estimates of POS, POD, and POC are as accurate as possible.

After re-evaluation, the planner must decide whether to proceed on the basis of new assumptions and scenarios, or to suspend active searching pending further developments.

Probable Success Rate (PSR)

Probable Success Rate (PSR) is the instantaneous rate of change in POS for adding one more increment of effort (one more searcher) to a search segment. It contains most of the elements of both POC and POD. The PSR for any given segment can be determined by the following equation:

$$PSR_{before} = \frac{(W) \times (v) \times (POC_{before})}{A}$$

In this equation, W is the effective sweep width or "detectability index," v is the average search speed (velocity) of the searcher, POC is the probability of containment and A is the segment's physical area.

A segment's PSR value, standing alone, is not useful nor intuitive. However, when the PSR values for all segments are computed and compared, the one with the highest PSR value is the one where the greatest return on investment from the first additional increment of search effort can be produced in the shortest time.

If the segments are sorted in descending order of PSR values, the segments are listed in order of most productive to least productive in terms of expected search results. PSR helps search planners allocate search resources to maximize the likelihood of success during a search period and over several operational periods.

Section 5-7 explains the use of PSR in the ground searching environment.

Section 5-3: Developing the Search Area

Search Area Concepts

Search planners must make crucial decisions as to where to look for the search object. Establishing the search area ultimately requires search planners to draw lines on a map encompassing the area(s) to be searched.

A search area is that territory or piece of ground within which the search planner believes the search object is located. The search area can, and in many instances does, change based on new information brought to light during searching and investigation.

Determining an appropriate search area requires careful evaluation of all the available information:

- Too large an area may increase the time to locate the search object or cause searchers to be spread too thin, and may result in missing the search object; and
- Too small an area may decrease the likelihood the search object is in the search area, delay finding the object until after the search area is expanded, or result in not finding the search object at all.

The search planners' job is to determine a search area that contains the search object, taking into account known information, clues, assumptions about the search object, and plausible scenarios. Search planners can use the Lost Person Questionnaire (Appendix E) to gather information about a lost person, and a Search Urgency Determination Form (Appendix D) to help assess the urgency of a particular missing person incident.

The concepts of Theoretical Search Area, Statistical Search Area, and Subjective/ Deductive Search Area (discussed in this Section) will aid search planners in establishing the search area for a particular incident. These concepts are used in a "process of elimination" or "process of successive refinement" to eliminate as much of the Earth's surface as possible from consideration for the search. To begin, the search planners will require a location and time describing where and when the search object was last known to be safe.

Last Known Position (LKP)/Point Last Seen (PLS)

The Last Known Position (LKP) is the last position based on clues or evidence belonging to the missing search object. LKP is similar to Point Last Seen (PLS), which is the location where the search object was last physically sighted. LKP/PLS is the starting point for determining where to search and the size of the intended search area. An accurate LKP/PLS is important to the search planner because:

- Having an accurate LKP/PLS will help to determine locations that may not require searching;
- In general, the sooner the LKP/PLS is known, the search area location determined, and the search commenced, the closer the search object will be to the LKP/PLS;
- If the LKP/PLS is closer to where the search object is believed to be located, then the search area can be smaller and more precise, leading to a more focused search (increasing the effectiveness and efficiency of the search, and reduces the amount of time it will take to locate the search object);
- The LKP/PLS may change as information is received during the search; and
- Statistical analysis of the search object's distance traveled is based on the initial LKP/PLS and where the search object is located.

Remember that even if the search planner is reasonably certain of the search object's LKP/PLS, an investigation into the events surrounding the disappearance of the search object should continue in full force until the search object is found or until active searching is suspended pending further developments. Sometimes clues found on scene will help establish a new LKP/PLS. However, updated information that may revise the LKP/PLS may also come from off-scene sources through investigation.

Theoretical Search Area (TSA)

The theoretical search area (TSA) is the maximum possible area within which the search object might be found. TSA is generally reported as a radius from the LKP/PLS (Figure 4-3-1). The radius represents the maximum possible distance the search object could have traveled under the expected conditions in any direction in the time since missing.

TSA can be used to determine the outer boundary of the search area, based on the available information. However, if considerable time has elapsed, the TSA can be very large. Given that searches frequently have limited search resources, the TSA is likely to be too large to search in the time available. Determining the TSA does help the search planner by finding a maximum distance the search object could have traveled in the available time. That maximum distance and area can be used in the investigation relating to the search. For the search planner, TSA:

- Provides an area to target when using the media to solicit clues;
- Helps evaluate clues (inside or outside the TSA);
- Determines an "area of interest" for investigation; and

- Encourages search planners to begin to consider areas within the TSA where it is more reasonable for the search object to be located and areas that are unreasonable.

Determining TSA Size

The following formulas can be used to help determine the size of the TSA for an overdue aircraft and a missing person.

TSA: Overdue Aircraft

To determine TSA radius for an overdue aircraft, several factors are required:

- Amount of fuel on board;
- Fuel burn rate (or rate of consumption);
- Aircraft speed; and
- Time to fuel exhaustion.

The formula for the TSA radius of an overdue aircraft is:

$$\text{Distance} = \frac{\text{Fuel X Cruise Speed}}{\text{Burn Rate}}$$

TSA: Missing Person

The TSA radius for missing persons is based on search object speed of movement and time elapsed since determining the LKP/PLS.

$$\text{Distance} = \text{Speed X Time}$$

Table 5-3-1 includes estimated missing person speeds. Some questions for the search planner to consider include:

- How many hours would the missing person walk in a day?
- Would the missing persons' motivation or physical condition change their speed?

Statistical Search Area (SSA)

The statistical search area (SSA) is based on statistical analysis of data on *where* similar search objects in similar conditions have been found in the past.

Historical data on where search objects have been found may come in different forms. For example, there is data on how far search objects have been found from the initial LKP/PLS. Depending on the source, the historical data may be quite general, taking in all types of terrain and search object/activity types, or the data may be broken down by terrain, search object, and activity types.

Historical data may also exist for specific areas, such as parks or specific regions within parks. Such data may be even more specific to particular trails, peaks, and/or other features.

The SSA provides a good starting point for determining an initial search area.

Statistical information on lost person behavior and find locations has been published by several organizations and individuals, and is readily available. Information on aircraft crash locations relative to intended routes have been published in studies by Canada, CAP, and most recently a study performed on behalf of NASA.

Use of statistical analysis is based on an assumption the search object(s) will act in a manner similar to those in the same types of situations. However, it is important to remember that determining the search area based on statistical analysis, although a sound initial practice, will not yield a search area that accounts for all the information and assumptions pertinent to a specific search.

However, determining an SSA is a good starting point, allowing the SAR planner to:

- Allocate limited resources to the most likely spots first;
- Target SAR resources to areas for initial confinement efforts; and
- Define an area the search planner would like to search given enough SAR resources.

As the search effort and investigation progress, the initial search area should be modified to more accurately fit the circumstances of the current situation. Since the existing historic data on searches and finds is not likely to be specific to the environment or object of a specific search, SAR agencies should consider collecting data on where search objects have been located in their areas of responsibility. Organizations can then use the data during future searches to assist in determining a more accurate search area.

There is one important difference between the TSA and the SSA that bears mentioning. The TSA continues to grow with the passage of time. The SSA is based on historical data about where search objects were found, usually without regard to when they were found. In other words, the way historical data is typically used for determining an SSA does not depend on how much time has elapsed between the initial LKP/PLS and the time when the search object was found. For this reason, it is possible, although unusual, for the TSA to be smaller than the SSA. In order for this to occur, the elapsed time between the initial LKP/PLS and the commence search time would have to be relatively short.

Subjective-Deductive Search Area (SDSA)

A subjective-deductive search area (SDSA) is based on an intuitive approach to determine the search area balanced by logic. The SDSA is the search area where SAR planners think, feel, and believe the object they are searching for is most likely located, given all available information for the specific incident, search object, and environment. Assessing all available information to determine an SDSA is necessary because:

- The TSA is usually too large to be effectively searched;
- The SSA is based on other cases that may or may not accurately portray the current search object and conditions;
- Modifying these search areas, based on data about the specific incident should help define a smaller search area that has an increased likelihood of containing the search object; and the
- SDSA size is based on likely scenarios and useful for setting search priorities.

The SDSA is used to modify the TSA and SSA to help determine the territory to search and how much effort should be allocated to each portion of that territory. It also helps:

- Define specific locations and routes to search;
- Select specific confinement locations and tactics; and
- Suggest alternative scenarios to be developed, investigated, and perhaps searched.

Table 5-3-1 and Figure 5-3-1 provide a comparison of TSA, SSA and SDSA.

Table 5-3-1: Comparison of Theoretical, Statistical, and Subjective-Deductive Search Areas		
Search Area	**Definition**	**Amplifying Information**
Theoretical Search Area (TSA)	Distance in all directions that the search object could have traveled based upon the amount of time since the subject was last seen.	**Estimated Theoretical speeds:** • Hiker No pack: 2.0-2.5 mph; • Hiker W/ pack: 1.5-2.0 mph; • Hiker uphill: + 1 hr/1000 ft Mountain; and • Child 3-4 (2.0 mph), 5-6 (2.4 mph), 7-8 (2.7 mph), 9-10 (2.9 mph), 11-12 (3.0 mph).
Statistical Search Area (SSA)	Statistical maximum zone (based upon 95%) derived from previous incidents of same type of search object. Refer to lost person behavior statistics available either locally or commercially for distances from initial LKP/PLS. Elevation and dispersion angle data, if available, can also limit the SSA.	It is also useful to plot the SSA's median or 50% ring on the map. However, the 50% distance must not be used for the search area boundary. It is possible for the theoretical to be smaller than the statistical distance. If sufficient containment exists then the search area can be kept smaller.
Subjective/ Deductive Search Area (SDSA)	Further limit the size of the search area based upon geographic and manmade features. Process of looking at all facts, scenarios, individual knowledge of the search object, unique factors of incident, previous history of the area, lost person behavior, clues, and investigative information to determine most likely areas.	Cliffs, rivers, major bodies of water, roads, and trails may be used to further limit the search area. However, keep in mind some search objects may cross roads (e.g., night, confused, dementia, etc.). The deductive part does not physically limit the search area but it does shift probability to more likely areas and helps to identify where to search.

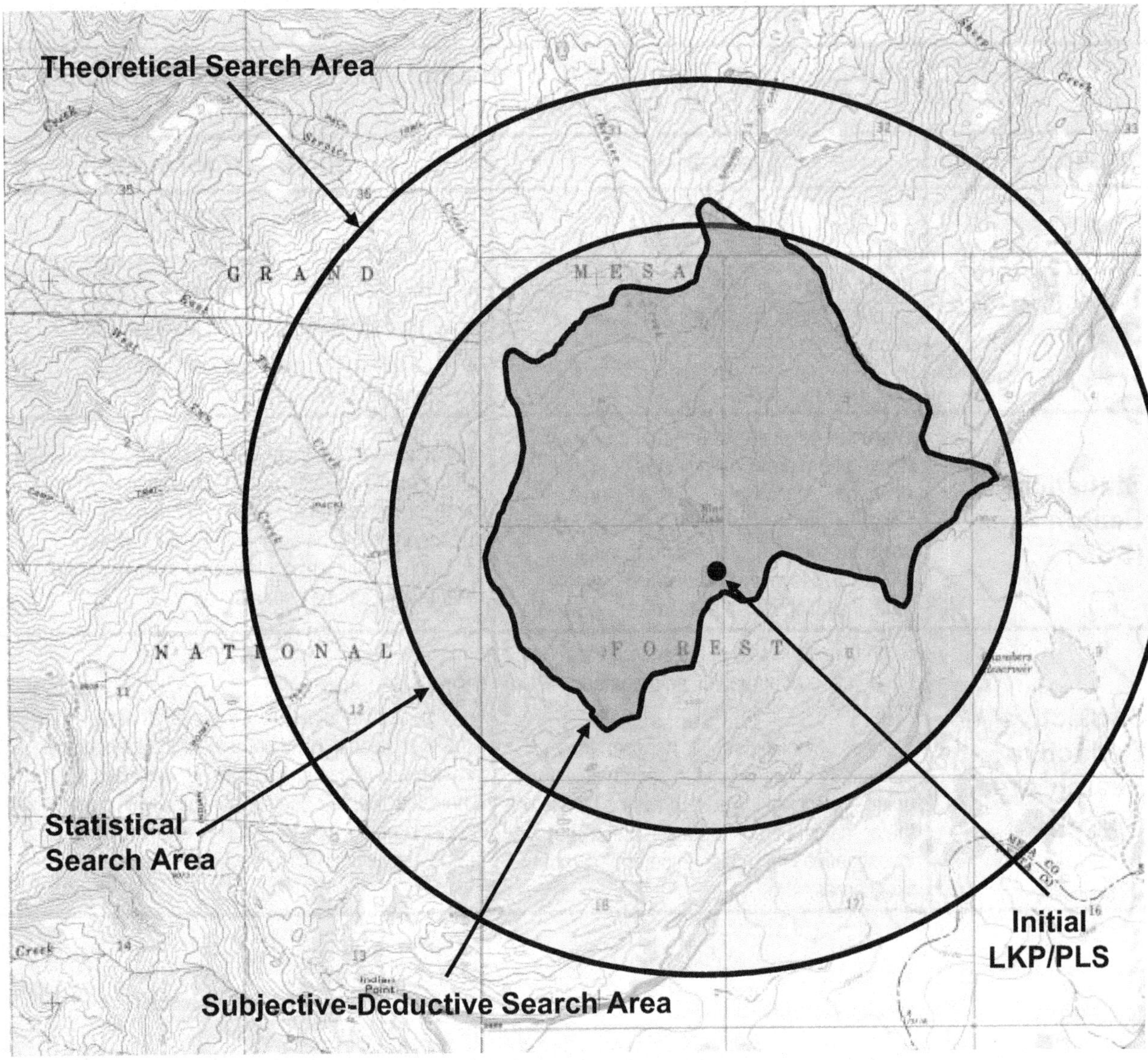

Figure 5-3-1: Relationship between Theoretical, Statistical, and Subjective/Deductive Search Areas (Missing Person Search).

General Scenarios

A scenario is a consistent set of known facts and assumptions describing what may have happened to the search object. The scenario usually consists of a sequence of actual and assumed events and is used as the basis for planning searches.

Search planners may be the source of some assumptions used to develop the scenario. There may very well be multiple scenarios that fit what is known about a case. To be useful, scenarios must be reasonable and consistent with known facts. Each scenario must be likely enough, given what is known about the search object, to warrant following-up with searching or investigation.

Any scenario that has a nontrivial probability of being valid in the view of search planners should be included in scenario analysis. Considering multiple scenarios can help the search planners:

- Integrate likely alternatives into a single search plan;
- Force search planners to consciously think about alternatives they had not considered on their own;
- Reduce common difficulties in developing POC consensus;
- Keeps each search planner from assigning POCs based on their individual and often contradictory or inconsistent scenario assumptions;
- Reduces difficulties with over and under estimating conditional probabilities;
- Helps set better priorities within the search area;
- Helps to identify better ways to confirm or refute various scenarios, leading to a better estimate of where the search object could be located;
- Helps to more precisely identify likely routes and locations of the missing persons, leading to quicker finds; and
- Better identify the types of clues that might be found, leading to better resource assignments.

Likely Route and Location Scenarios

Likely route and location scenarios are routes along which the search object likely traveled and the specific locations where the search object is likely to be found. These scenarios can form the basis for rapid search tasking and further search area discussions.

Choosing likely routes the search object may have traveled from the initial LKP/PLS to the current position will help find clues. As the search object traveled, it was likely influenced by the surroundings. There may have been visible, audible, desired, or imagined things which attracted the search object or dissuaded it from continuing. Perhaps there is a hazard that affected the search object's behavior. Identifying this route allows the search planner to predict where the search object may go and to assist in search planning.

This page intentionally left blank.

Section 5-4: Distress Beacon Search Concepts

Distress Beacon Response Policy

In response to distress beacon alerts, RCCs should consider all available information such as position, registration, and the presence of corroborating information. RCCs should evaluate distress beacon reports and attempt to correlate them with other indications of distress. Concurrently, they should attempt to obtain additional information on those involved. RCCs should expand their investigations as necessary to aggressively pursue the cause of alert signals and dispatch SAR resources to assist as circumstances require. Types of distress beacon alerts and response policy guidance are presented in Table 5-4-1.

Table 5-4-1: Distress Beacon Alert and Corresponding Emergency Phase	
Distress Beacon Alert	**Emergency Phase**
406 MHz registered and located alert (E solution* or Doppler) with location correlating to an area of known or possible distress.** **406 MHz unregistered and located alert (E solution* or Doppler**) with location correlating to an area of known or possible distress**	Initially evaluate as DISTRESS
121.5/243.0 MHz multiple reports of audible alert **406 MHz registered but un-located alert** **406 MHz unregistered and un-located alert** **406 MHz registered and located alert (E solution* or Doppler**)** **406 MHz LEO unregistered and located alert (E solution* or Doppler**)**	Initially evaluate as ALERT. Investigate, reevaluate, and respond as facts and circumstances warrant.
121.5/243.0 MHz first report of audible alert	Initially evaluate as UNCERTAINTY. Investigate, reevaluate, and respond as facts and circumstances warrant.
* "E" solution being a GPS position (GEO satellite). ** "Doppler" being a position provided by Cospas Sarsat LEO satellite.	

406 MHz Distress Beacon Cospas-Sarsat Alerts

Since 1990, distress beacon technology has been moving to a solely dedicated frequency for satellite distress beacons, 406 MHz. Use of this frequency will minimize interference problems. In addition, satellite software recognizes and relays only coded 406 MHz distress beacon signals, minimizing false alerts. Accordingly, response to 406 MHz distress beacon alerts is immediate, keeping in mind the precepts of risk management.

First alerts and composite solutions for 406 MHz distress beacons indicate a beacon has been activated. SAR response to a 406 MHz distress beacon alert should approximate the response to a MAYDAY call. The 406 MHz Cospas-Sarsat system and equipment yield high confidence alerts and positions. However, factors such as satellite pass geometry, atmospheric anomalies, and beacon oscillator stability may degrade the beacon signal and position data. Any alert degradation is usually reflected in the split between A and B solution probabilities on first alert messages.

Registered But Unlocated 406 MHz Alerts

Normally, registered but unlocated 406 MHz alerts are treated as a distress. RCCs should use all reasonable means to determine the position of the person in distress.

As registered but unlocated 406 MHz alerts contain no position information, RCCs must exploit all reasonable means to ascertain at least a general distress position. Armed with a general position or usual operating area and suitable homing capable response assets, RCCs can render timely, effective assistance.

Distress beacon registration points of contact are usually the most promising leads for information, particularly for position, situation, and further points of contact. When only general position information is available, suitable aircraft should be launched to direction find (DF) on the distress beacon's signal.

Unregistered/Registered and located GPS Protocol Distress Beacons

Location Protocol Beacons (or GPS Protocol Beacons) contain a GPS chip that can accurately calculate the position of the distress beacon and transmit that position as received by the satellite. Since the Cospas-Sarsat system requires multiple passes from low earth orbiting satellites to calculate the distress beacon's position by Doppler shift, this technology provides a more timely method of notifying SAR responders of a distress beacon's position.

For alerts that contain an encoded GPS position (described in alert messages as an "E" solution), responders shall evaluate it as a distress incident regardless of whether the distress beacon is registered or if a location has been determined by the Cospas-Sarsat system.

When a composite position is obtained by Cospas-Sarsat satellite passes, search planners should compare the encoded GPS position to the composite solution to verify the location of response.

50/50 Split Solutions

50/50 splits are no different than other A/B solutions and merely indicate that mathematically the distress beacon could be in either location. Plotting the position and carefully analyzing the distress beacon data and registration information will usually allow you to determine the actual location. Also note that 50/50 solutions tend to be less accurate than other solutions.

Alert Query Reports – USMCC Database

In pursuing amplifying distress information, RCCs may query the USMCC database to see whether or not a particular 406 MHz distress beacon has been activated, or to check for all distress beacon activations over a specific time period or in a specific area. RCCs may do so by requesting an Alert Query Report from the USMCC. The USMCC user's manual has more information on this process and guidelines for interpreting these reports.

Cospas-Sarsat System Limitations

As with any tool, search planners must be aware of the Cospas-Sarsat system's limitations. False alerts (e.g., inadvertent activations or hoaxes), non-distress beacon alerts, and unresolved distress beacon alerts reduce the efficiency of the Cospas-Sarsat system. Operator misuse, equipment malfunctions, improper testing, and hoaxes may downgrade distress beacon effectiveness.

However, the "A" solution on a 406 MHz first alert will be the correct position 95% of the time. SAR resources can reasonably be dispatched immediately upon receipt of a 406 MHz first alert. A composite solution will be received, provided the beacon continues to transmit, usually within 60 minutes (average time is approximately 45 to 90 minutes).

121.5/243.0 MHz Distress Beacon Alerts

The operation of EPIRBs on 121.5 MHz frequency became prohibited in the U.S. on January 1, 2007. This prohibition does not apply to Emergency Locator Transmitters (ELTs) normally carried on aircraft or on man overboard devices.

The International Cospas-Sarsat System terminated satellite processing of 121.5/243.0 MHz distress signals on February 1, 2009. Although RCCs no longer receive 121.5/243.0 MHz Cospas-Sarsat Alerts, they will still receive and should respond to audible 121.5/243.0 MHz alerts as indicated in Table 5-4-1 and outlined below.

Authorized 406 MHz distress beacons do contain a 121.5 MHz homing signal, which will continue to be used for DF.

121.5/243.0 MHz Distress Beacon Audible Alerts

The first report of an audible 121.5/243.0 MHz distress beacon alert corresponds, at a minimum, to the Uncertainty Emergency Phase. RCCs should aggressively attempt to corroborate 121.5/243.0 MHz audible alerts with any other potential distress information such as FAA issued INREQ, ALNOT, reports of overdue aircraft, or other distress reports.

A second audible report of a 121.5/243.0 MHz distress beacon alert corresponds to the Alert Emergency Phase. Although response to 121.5/243.0 MHz audible reports may represent a significant resource commitment with a limited likelihood of actual distress, once additional audible reports have been received, RCCs should make every effort to locate and determine the source of the signal. The RCC should evaluate the reports by plotting and monitoring individual reports, identifying search areas, and notifying corresponding FAA facilities, airfield control towers, fixed base operators, and state SAR Coordinators/Responsible Authorities, as applicable, to further identify a defined search area to determine if the signal is distress related. While it is common to determine that a signal source is emanating from an area where distress is unlikely (e.g., marina, anchorage, airfield, etc.), it should still be investigated so that the signal can be silenced.

When receiving audible reports of a 121.5/243.0 MHz distress beacon alert, RCCs should seek specific information from the reporting source, including:

- Reporting source location;
- Reporting source course, speed, and altitude; and
- Strength of the signal heard.

Search Areas Based on Audible Reports

Audible distress beacon alerts may not always indicate a distress, but they do indicate that a distress beacon has been activated. Historically, many of these alerts have been false alerts resulting from hard aircraft landings or caused by human error during aircraft or vessel maintenance.

121.5/243.0 MHz distress beacon audible alerts are normally reported by airborne aircraft and FAA facilities, and will usually help in determining the distress beacon's location.

Pilots are encouraged to monitor 121.5 or 243.0 MHz during flight and report hearing audible ELT signals to the appropriate FAA facility. The report should include the time, position and altitude at which the signal was heard (i.e. 1530z; Norfolk (ORF) 270/30; ALT: 11000).

It is important to gather reports from different locations and the lowest possible altitudes in order to narrow down the search area. Follow up with appropriate FAA facilities, airports, and other necessary outside agencies for additional leads.

Radio signals have a "horizon" beyond which they cannot reliably be heard. This distance varies with the altitude of the receiver. See Table 5-4-3.

Knowing the position and altitude at which various aircrews hear an ELT signal, the search planner can draw line-of-sight (LOS) circles to help narrow down the likely location of the ELT. It also gives searchers a position to allocate aeronautical assets when initiating a search.

In the example below (Figure 5-4-1), the ELT is most likely within the relatively small area where all three circles overlap. If an aircrew does not hear the signal, the ELT would then likely be in the area not overlapped by that circle; however, there are multiple reasons why a signal might not be heard. Gathering additional reports will allow the search planner to refine the search area and possibly determine the signal's location. In every case, the signal should be investigated as long as the distress beacon's signal continues. If SAR resources are unable to locate the signal, the last resort should be contacting the FCC for assistance.

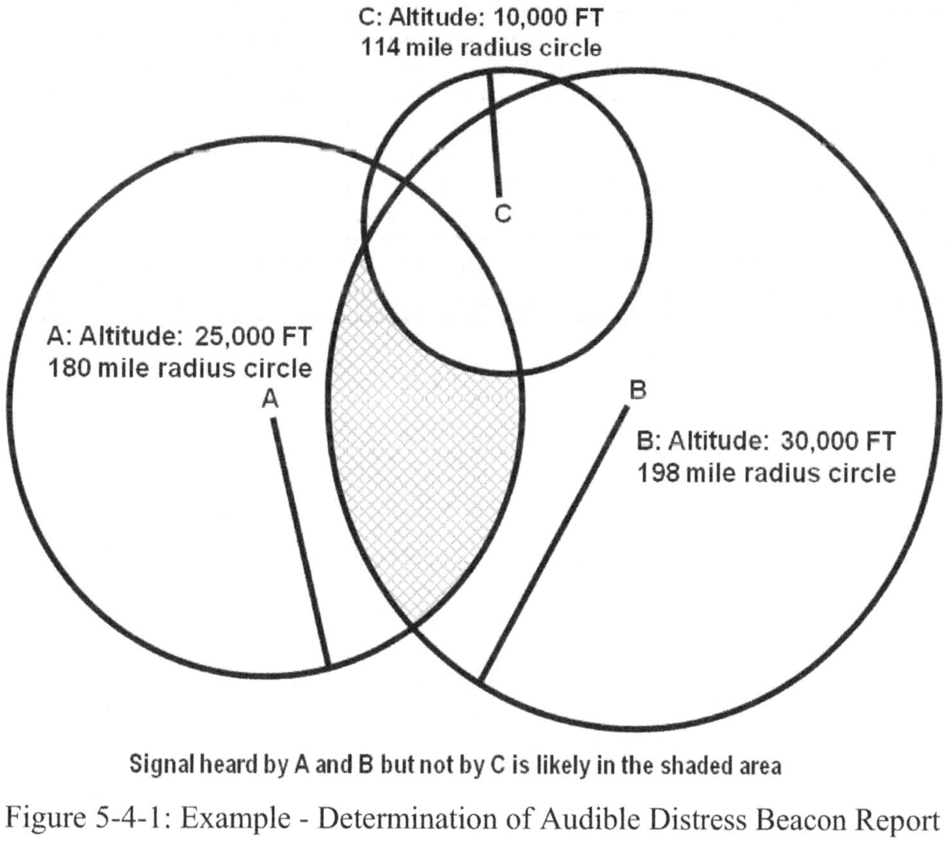

Figure 5-4-1: Example - Determination of Audible Distress Beacon Report

Direction Finding (DF)

DF radio receivers can indicate the direction from which a signal is emanating and can be invaluable in finding and isolating distress beacons. Three examples of particularly important DF uses include the following:

- Selecting the right aircraft on a crowded airfield;

- Finding distress beacons in non-aircraft locations (e.g., hangar floors, cars, etc.); and
- Finding downed aircraft in unpopulated or sparsely populated regions.

Triangulation (Build-and-Fade Search)

By gathering the correct information from pilots, searchers can narrow the search area very quickly. This technique also works for ground-based receivers.

It is important for search leaders to be able to direct searchers in this procedure and to plot and make good use of the information generated. Table 5-4-2 below provides suggested steps for determining the location of an ELT signal. Figures 5-4-2 and 5-4-3 below graphically demonstrate the procedures outlined in Table 5-4-2.

Table 5-4-2: Determining Location of ELT Signal	
Step 1	Aircrew hears an ELT signal on 121.5 and/or 243.0 MHz
Step 2	Crew adjusts squelch or volume so that the ELT signal is barely audible, notes their location as "first heard," and continues to fly in a straight line
Step 3	The signal should become stronger & louder. If not, the crew reverses course to reacquire the signal
Step 4	If possible to detect, the crew notes the aircraft location where the signal is the strongest/loudest
Step 5	The crew notes the location where the ELT is no longer audible (last heard)
Step 6	Halfway between the first and last heard points should be their point of closest approach. The ELT should be directly off the wing; however, the crew cannot yet determine which wing, left or right
Step 7	Repeating these steps along a different course or with a different aircraft can allow searchers to determine the approximate location of the ELT signal. Additional passes should help narrow the solution

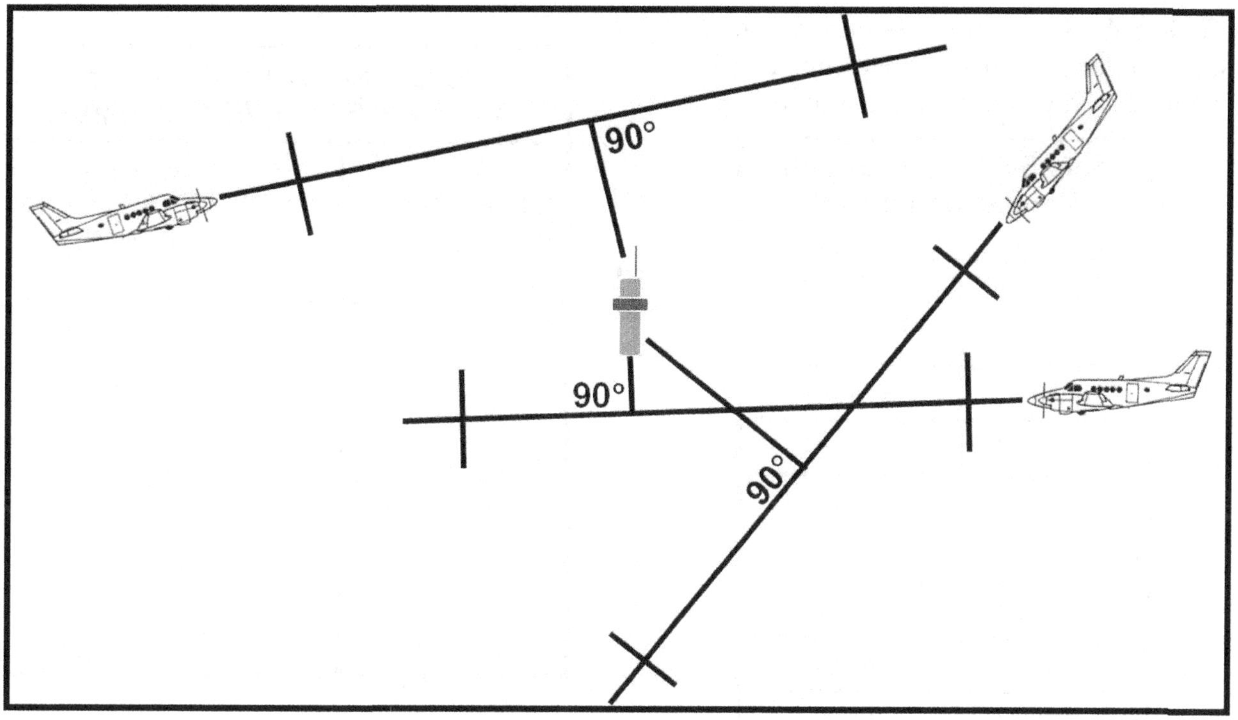

Figure 5-4-2: Example #1- Aircraft ELT Signal DF

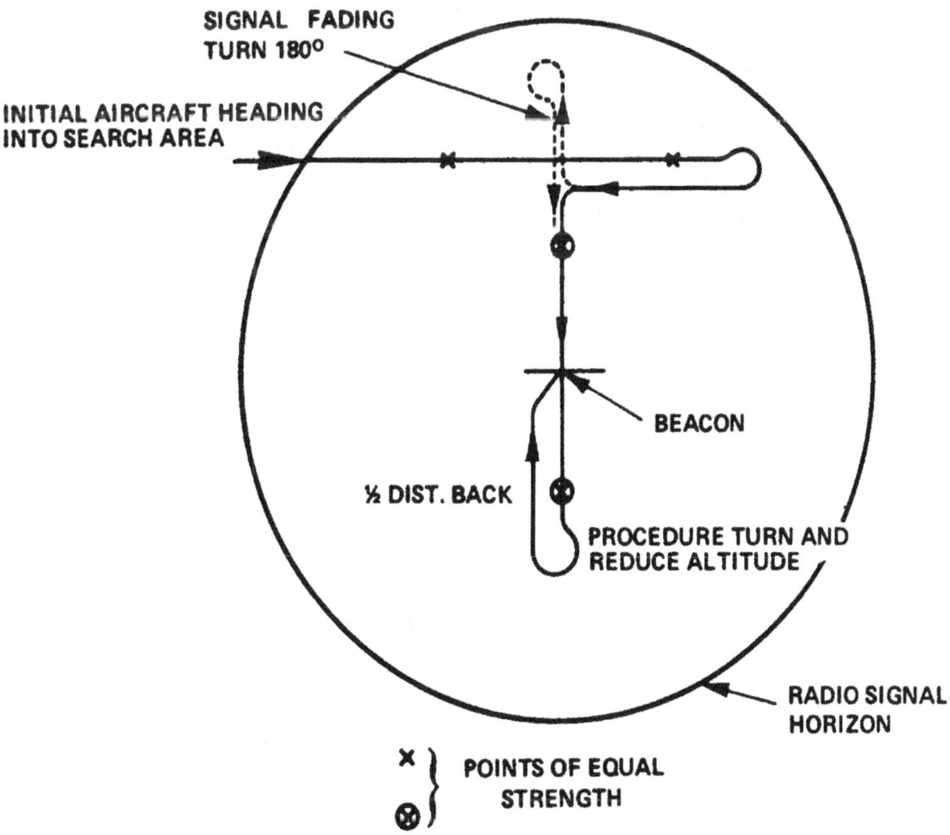

Figure 5-4-3: Example #2 - Aircraft ELT Signal DF

Table 5-4-3 below provides line of sight distance for aircraft when detecting an ELT signal, based on the aircraft's altitude.

Table 5-4-3: Line of Sight Values for Aircraft
(Altitude vs. Line-of-Sight)

Altitude	Line-of-Sight
1,000 FT	36 miles
2,000 FT	51 miles
3,000 FT	63 miles
4,000 FT	72 miles
5,000 FT	81 miles
10,000 FT	114 miles
15,000 FT	140 miles
20,000 FT	162 miles
25,000 FT	181 miles
30,000 FT	198 miles
35,000 FT	214 miles
Ref: FRCC and Canadian S R calculations. alues are theoretical and can be affected by signal strength, weather, terrain, condition of the beacon, and other factors.	

Other DF Techniques

Table 5-4-4 provides procedures for two ways to locate a strong ELT signal that is overpowering the receiver, or a signal that could be coming from one of several near-by aircraft:

Table 5-4-4: Procedures for Locating Strong ELT Signals

De-sensitize the receiver	1. Remove or collapse the receiver antenna. 2. De-tune the receiver slightly off of 121.5MHz.
Body Shadowing	1. Hold the receiver close to the chest and turn around slowly, using the body to partially block the signal. 2. The signal should be coming from the direction of best/loudest reception; alternatively, it is coming from behind when the signal is weakest (effectively blocked by the body).

Additionally, a distress beacon signal that has been activated may be located even without a special receiver. When using a tunable FM radio, the following procedures are suggested:

1. Tune a tunable FM radio to 100.1 MHz and use body shadowing; and
2. As the signal gets stronger, de-tune farther down the scale.

Note: This technique is based on the FM radio's image frequency principle, so older-style manual tuning radios work best. Newer, expensive radios filter their signals better and will likely not work.

Continued Monitoring

The SMC/IC will obtain valid and useful information when determining the source of distress beacon signals based on airborne and ground reports received. The SMC/IC should thoroughly investigate and evaluate the data to determine the signal's location and correlate the signals against any active INREQ, ALNOT, or other indications of potential or actual distress. Consider all reports as originating from a distress situation until proven otherwise.

Additional airborne and/or ground reports of 121.5 MHz audible alerts or lack of ambiguity with known reports may move a case to the Alert or Distress Phases. The distress Phase can be equated to opening a RCC mission. Each additional report will enable the search area to be refined to better identify boundaries and responsible agencies. The RCC will alert appropriate State Agencies per SAR agreements and/or SAR unit(s) for distress beacons that cannot be positively identified as a false alert, or readily located and silenced.

The RCC retains the operational direction for cases involving high altitude reports that span multiple jurisdictions, unless such jurisdictions deem it necessary for RCC notification involving all alerts that fall within their boundaries regardless of altitude.

State Policy

Effective response to the activation of a distress beacon requires a good relationship between the State where the distress beacon is detected and the RCC.

States must develop internal procedures, reporting requirements, and the assignment of responsibilities when a distress beacon signal is detected. Once these practices are established, it is imperative to update the memorandum of understanding between the State and the RCC to ensure proper protocols are followed.

When a Distress Beacon Is Located

Distress beacon signals may mask additional actual distress signals and so must be located and turned off.

CAUTION: SAR authorities have no special entry authority in non-distress situations and the rules of trespass still apply, meaning the permission of the property owner typically will be required in order to enter private property, including private vehicles. Sometimes, getting local law enforcement involved can help to secure permission or to obtain authority to enter the property.

The first action should usually be to notify the owner to either shut off the distress beacon or to seek permission to assist. Take care to not damage the radio, aircraft, or other property.

As a last resort, consider advising the owner that federal laws provide for monetary penalties of up to $16,000 for each such violation or each day of a continuing violation and up to $112,500 for any single act or failure to act for non-distress activation of a distress beacon or knowingly transmitting a false alert or hoax distress signal. Fully document and report to the RCC or FCC any such cases.

Anytime SAR responders work on or disables a distress beacon, leave a warning note in a prominent place for the owner and contact the nearest FSS with time of deactivation, tail number, owner's name, beacon make/model, and circumstances for deactivation.

Report Location of the Distress Beacon

Report all pertinent information to the SMC/RCC so the case can be closed and the reporting source notified. Information should include: the location (name and coordinates), time located, time silenced, beacon manufacturer model and serial number, battery expiration date, owner information, aircraft type, and tail number.

Further monitoring

The RCC should continue to monitor 406 MHz signals through the Cospas-Sarsat system, which predicts future satellite pass times and identifies when a distress beacon signal was expected, but not heard. The

RCC will generally monitor the case until the satellite no longer hears the distress beacon signal for at least two satellite passes before closing the mission.

For signals only heard on frequencies 121.5 MHz or 243.0 MHz, the RCC will verify the signal has been silenced by requesting the FAA solicit further pilot reports to determine the signal is no longer being heard before closing a mission.

The additional monitoring should:

- Detect whether the silenced distress beacon was covering up another potential distress signal; and
- Detect whether the initial distress beacon is reactivated (perhaps through handling after silencing).

All distress beacon signals must be treated as an actual distress situation until confirmed otherwise. False alerts can cover up an actual distress situation and must be located and silenced quickly.

PLBs and other Devices

Personal Locator Beacons (PLBs) are portable distress beacons that operate much the same as EPIRBs or ELTs. These distress beacons are designed to be carried by an individual instead of on a boat or aircraft. Unlike ELTs and most EPIRBs (which may be activated both automatically and manually), PLBs can only be activated manually and operate exclusively on 406 MHz. All U.S. coded distress beacons also have a built-in, low-power homing beacon that transmits on 121.5 MHz. This allows SAR resources to DF on a distress beacon once the satellite system provides position information.

Distress beacons will always be most effective when combined with a visual and/or audible distress signal such as a mirror, whistle, or a strobe light to help catch the attention of SAR resources.

Concerning PLBs:

- While there is no licensing requirement, registration with NOAA is mandatory under FCC rules once the PLB is purchased;
- Active PLBs are monitored at the RCC, who passes the information to the appropriate State agency; and
- The distress beacon requires a two-step manual process to be activated.

(NOTE: There are commercial enterprises that have begun offering PLB type "services" associated with non-PLB devices. These devices do not alert through the Cospas-Sarsat system and do not have to abide by Federal PLB procedures, guidelines, or standards. While these devices offer similar services to PLBs, they are not considered PLBs under FCC standards and are outside the national SAR infrastructure. Responsible SAR authorities may receive distress notification from commercial monitoring services for these types of devices. More information on these types of beacons is available from the manufacturers.)

Section 5-5: Extended Search Planning Concepts (General)

Searches

Consensus in Search Planning

Proportionality

Searches

Initial search operations locate the majority of search objects. However, failure to recognize the potential need and plan for extended operations jeopardizes search safety, efficient operations, and can delay locating the search object.

Extended search operations involve:

- Determining if the SMC/IC transition is necessary;
- Determining formal operational periods;
- Establishing organizational structure;
- Establishing incident objectives;
- Formally determining the search area;
- Regionalizing and segmenting the search area;
- Establishing regional probabilities of containment using a consensus method;
- Determining tactics required to implement strategies set from objectives;
- Prepare and conduct planning meetings;
- Determining resource requirements;
- Preparing and approving an Incident Action Plan (IAP)/SAR Action Plan (SAP) to include allocation and tasking of resources;
- Ensuring that personnel are rested;
- Ensuring smooth personnel transition;
- Providing operational briefings;
- Deploying resources;
- Tracking, evaluating, and following-up on clues and investigative leads;
- Planning for the next operational period;
- Debriefing field resources;
- Evaluating search results;
- Revising POC for searched areas;
- Projecting possible POD from effective sweep width tables;
- Reassessing priority of where to deploy resources;
- Continuing to execute the search action plan and assess progress; and
- Repeating the operational period planning cycle (Figure 5-5-1).

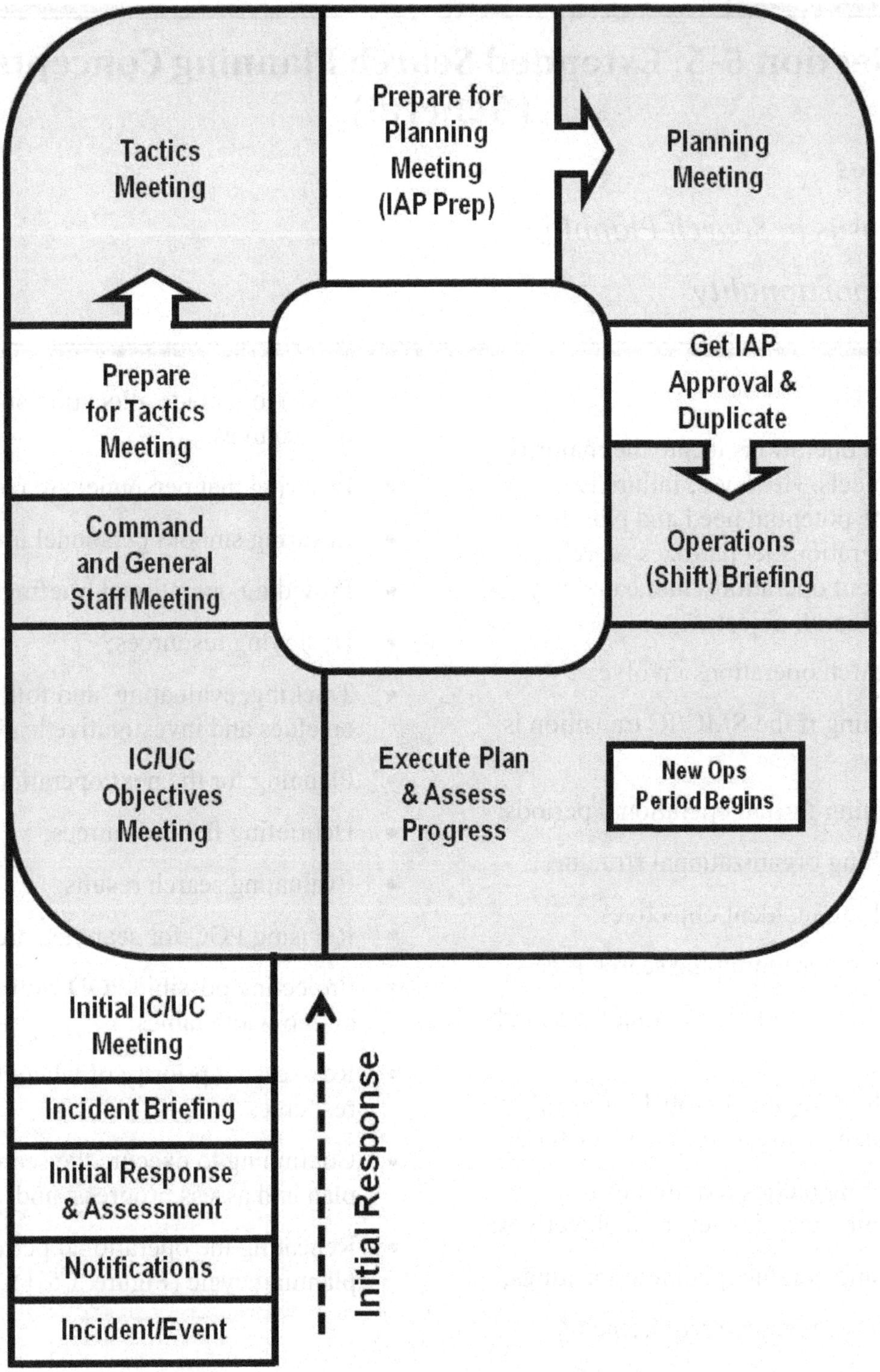

Figure 5-5-1: Planning "P" Operations Planning Cycle

Consensus in Search Planning

Consensus is a general accord or agreement. In search planning, this agreement relates to where the search planners believe the search object is likely to be located.

The consensus process is used to obtain input from expert individuals and aggregate opposing judgments in order to make more fully informed and considered decisions.

At many junctures during a search mission, it is desirable to collect the experience, judgments, and expert opinions of the search leadership. Consensus among the experts can help:

- Ensure that all important perspectives (including opposing views) are considered and that none are lost or ignored;
- Ease the conflicts that can arise over strongly held opinions;
- Build trust and confidence in the search plan and the leadership team; and
- Design a plan that reflects the participants' combined intuition and expert knowledge.

The consensus process allows the search leadership team to apply their experience, judgment, and knowledge of the situation to assign priorities to various parts of the search area. By involving multiple experts, consensus helps avoid individual bias and reduces "channelized vision." Consensus can be beneficially used when combined opinions and judgments may help:

- Develop and weigh scenarios;
- Develop regions;
- Establish/revise POC; and/or
- Evaluate clues and how they affect the search.

Proportionality

Proportionality is a set of numbers reflecting the overall leadership team's combined view of the relative likelihood of, or preference for, the available search options. These numbers are percentages; for instance, a team might believe there is a 50% probability the missing person followed the creek, a 30% chance of following the trail, and a 20% chance of not following.

No matter the method used (e.g., Modified Mattson Consensus, Mattson Consensus, Proportional Consensus, etc.), the results must accurately reflect the proportion, ratio, and relative amount of probability to be assigned to the various search options.

This page intentionally left blank.

Section 5-6: Extended Missing Aircraft Search Concepts

Searching for Missing Aircraft

Computer-Based Search Planning Aids

Searching for Missing Aircraft

Searching for missing aircraft is decidedly different than missing person searches. Planning an air search for an overdue or missing aircraft initially involves estimating the most probable location of a distress incident or of the incident's survivors.

One of the initial challenges in searching for a missing aircraft is determining the size of the search area large enough to ensure that the aircraft is somewhere in the area, yet small enough to be able to search with the available SAR resources. Second, search planners must choose SAR resources for the search and the search patterns to be employed to effectively cover the area.

Some aircraft searches are relatively simple if enough evidence such as eye witness reports, distress beacon coordinates, or radar data points provide the location of the downed aircraft. The position of a distress incident can be determined within fairly narrow limits if the following data is available:

- Location where the aircraft disappeared off radar (a 2009 NASA study showed that 50% of missing aircraft with radar data available were found within 0.8 miles of the last radar return[1]);
- Bearing or fix provided by radar, ground station or emergency radio aid;
- Dead reckoning position based on time of LKP/PLS; and/or
- Reports of sightings.

Radar data is probably the most important piece of information to help narrow down the search area. RCCs are the primary agency to assist in gathering radar data from the FAA as well as other information (see Section 1-4 - RCC Information Sources).

However, many cases may not have an exact location available, so determination of a search area assumes a very different character. Applying the basics of search area determination (e.g. TSA, SSA, SDSA), and using all the evidence from the investigation, helps narrow the search area. When detailed information is not available, the probable location of an aircraft may be narrowed considerably if information about the flight in question and other information is available, such as:

- A flight plan or aircraft route;
- Weather information;
- Aircraft audio information;
- Proximity of airports along the track;
- Pilot habits, record, experience; and
- Any other clues as to the location of the aircraft.

[1] Robert J. Koester, "Data Collection and Analysis for NASA World Wind Search and Rescue Visualization Program" (2009).

When little information is available, use of statistics can help define the search area. Search planners can use statistics from a 2009 NASA study to determine POC, POD, and POS for missing/overdue aircraft. These statistics help define the Air Search Area Definition (ASAD). Training and information on how to define search areas for aeronautical searches using statistics is available from the AFRCC (Basic Inland SAR Course) and the National SAR School (Inland SAR Planning Course).

Computer-Based Search Planning Aids

Computer-based search planning aids can be used to relieve the search planner of much of the computational burden involved in planning searches.

While in its infant stages, computer programs are beginning to provide modeling for statistics in both the air and ground environments based on new and in-depth research. Programs are constantly being developed and modified to model aircraft and lost person behavior. These programs can be used to calculate sweep widths, search endurances, search efforts, search areas, coverage factors, etc., as well as compute probability of success estimates and optimal allocation of available search effort. Additionally, as programs advance, they may be able to display, compare, and combine clues and probability maps associated with different scenarios, environmental/weather considerations, search patterns, track spacing, turn points for each leg, and time to complete searches at various search speeds.

Remember:
Computers and computer-based search planning aids do have limitations. They are tools to enhance and aid, but do not replace search planner judgment, intuition, and experience.

Section 5-7: Extended Ground Search Concepts

Effort Allocation

Probable Success Rate (PSR)

Optimal Allocation

PSR Limitations

Adjusting Search Calculations

Effort Allocation

The single factor over which search planners have the most control is the number of searcher-hours they allocate to the search segments. Unfortunately, searcher-hours can be a difficult number to track if the various search segments take more or less time to search. Therefore the required equations have been modified to allow the search planner to allocate "searchers" into the search segments. However focusing on searchers instead of searcher hours can occasionally be misleading.

In all effort allocation assignments, a key and essential assumption is that each segment searched is searched equally well (uniformly) throughout:

- Searchers do not miss chunks of a segment, including edges and hollows; and
- Searchers searched all portions of a segment.

Probable Success Rate (PSR)

Probable Success Rate (PSR) is the instantaneous rate of change in POS for adding one more increment of effort (e.g., one more searcher) to a search segment. It contains most of the elements of both POC and POD.

PSR is calculated as per the following formula:

When allocating searchers as the unit of effort, the Probable Success Rate (PSR) for each segment is:

$$\text{PSR} = \frac{(\text{search speed}) \times (\text{sweep width}) \times (\text{segment POC})}{(\text{segment area})}$$

or,

$$\text{PSR} = \frac{(vW) \times (\text{segment POC})}{(\text{segment area})}$$

Optimal Allocation

Scientists have found that the best distribution of effort is the one that "levels" or "evens out" PSR as much as possible using the available effort. This is illustrated below in Figure 5-7-1 with a three-dimensional bar graph that displays PSR values as the heights of columns, each column representing a search segment. In the Figure, a total of 15 search segments are shown. The four segments shown in green are tied for "first place" in terms of highest PSR. Three segments are tied for "second place" in terms of PSR (shown in orange), followed by the six segments tied for "third place" (shown in magenta), and bringing up the rear in "fourth place" are the two lowest PSR segments (shown in blue).

The principle works on the idea that as the level of available effort is allocated into the four search segments with the highest PSR (green in this example), the highest OPOS results from putting just the right amount of that effort in each of the highest-PSR segments so that all four PSR_{after} values fall together in lock-step and remain equal. This remains true until the four segments that initially had the highest PSR value have their PSR_{after} values reduced to the PSR level of the segments that initially had the second highest PSR values (shown in orange).

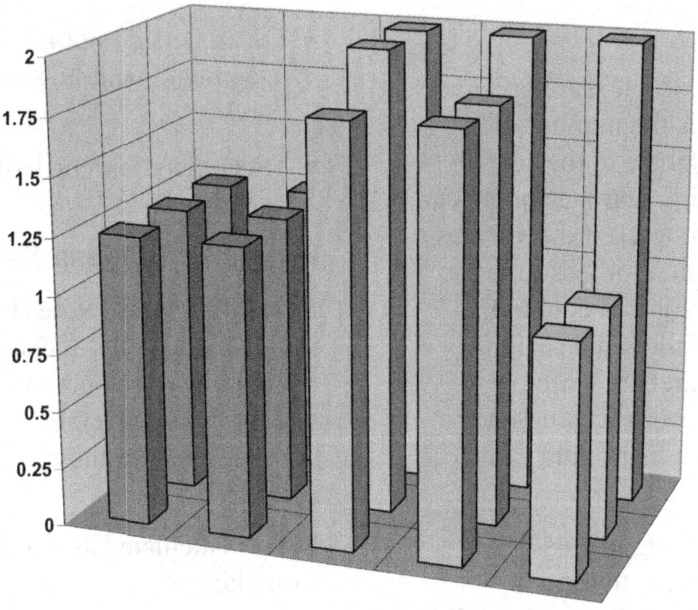

Figure 5-7-1: Initial PSR Values

Figure 5-7-2 shows the situation once the four segments that initially had the highest PSR values have been searched just enough so their adjusted PSR_{after} values have been reduced to the same level as the three segments that initially had the second-highest PSR values.

At this point, the process is repeated, except that there are now seven segments tied for highest PSR (orange and green). Therefore, as more effort becomes available, it is allocated to these seven segments so their adjusted PSR values are driven down by searching in lock-step until all seven segments (green and orange) have adjusted PSR values equal to the PSR values of the six segments (magenta) that were initially in third place.

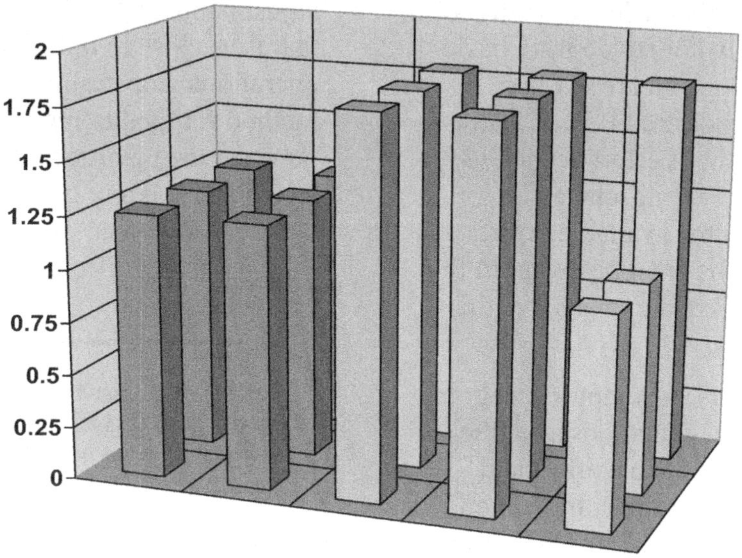

Figure 5-7-2: Adjusted PSR Values
(After optimal search of top four segments)

Figure 5-7-3 below shows the situation after the top seven segments have had their PSR values reduced to the same level as the six segments that initially had the third-highest PSR values. Whatever additional effort becomes available, it should be spread across the thirteen segments that are all tied for highest PSR. The search should continue in a like manner until all segments have the same PSR value. Thereafter, all segments should be searched so their adjusted PSR values continue to decrease in lock-step.

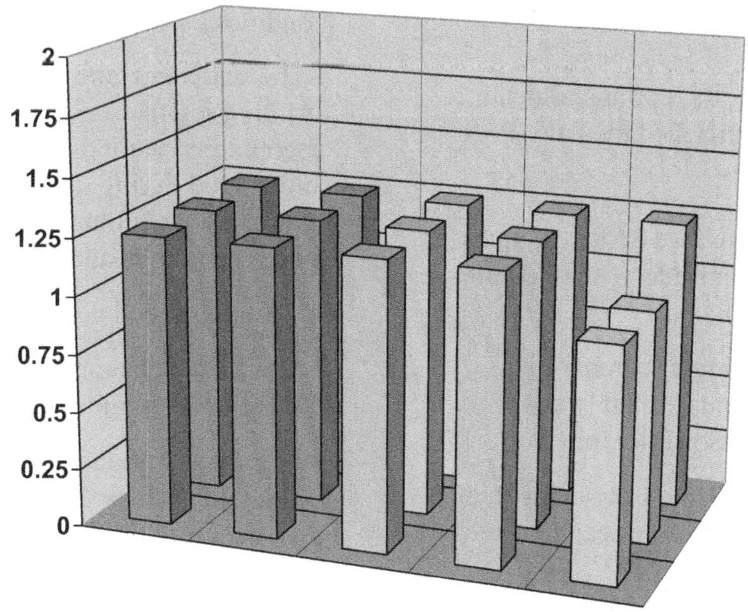

Figure 5-7-3: Adjusted PSR Values
(After optimal search of top seven segments)

So far we have described what needs to be done, but not exactly how to do it. In basic terms, the equations given are worked "backwards" from a desired PSR_{after} value to determine the amount of effort needed to achieve the desired PSR. If total effort needed is more than the available effort, then the computations are repeated until the available effort is adequate to achieve the desired PSR_{after}.

An alternative is to use a computer program that completes the computations after the user enters the total available effort and other inputs required for the computation (e.g., the effective sweep width (W) for each segment, the average search speed (v) in each segment, the POC for each segment, and the physical area (A) of each segment). The program then shows how much of the available effort should be allocated to each segment being searched.

The Inland SAR Planning Course at the National SAR School currently uses such a program, which is available through the School website:

http://www.uscg.mil/tcyorktown/ops/sar/inland/default.asp.

As discussed above, the optimal allocation of effort can be approached by using PSR.

PSR Limitations

The PSR method assumes SAR resources can be divided into infinitely small chunks, which could result in:

- Assigning one single searcher to a particular segment (probably not operationally reasonable); or
- Part of a searcher into one segment, and another part of the same searcher into another segment.

However, given a theoretically optimal allocation as a starting point, the search planner can make the necessary operational adjustments and still provide an allocation that is as close to optimal as possible for the operational constraints. This also provides a method for evaluating the difference between the theoretical "optimum" allocation and any other allocation. The search planner can estimate just how much POS is lost by different allocations of resources.

> No mathematical method can be allowed to take the place of good judgment in the field. The mathematics in this Addendum provides valuable decision aids, but cannot make decisions; mathematics only processes the available data and may not account for important, operationally significant realities.
>
> On the other hand, while pure intuition is easier to use, it is harder to justify later if success does not come early and it is not as reliable in the long run. Use the mathematics as a guide, but not as the complete answer.

Adjusting Search Calculations

Adjusting for actual conditions after the search is a vital step. The searchers must let the search planners know the actual conditions encountered.

If the searchers take less (or more) time to search a segment, obviously the conditions encountered were different than what was anticipated during search planning. The search planners must discover what changed and adjust the results accordingly if:

- Search speed was different;
- Searchers changed spacing but searched the entire segment; or
- Searchers searched only part of the segment (may also have changed spacing).

Remember that the methods presented are for area/grid searches, not for rapid (hasty) searches.

It is assumed that good search sense has led the search leaders to apply suitable rapid (hasty) search techniques to follow up on likely routes of travel, attractions, hazards, etc., along with confinement and other strategies as appropriate. However, the mathematical model supports rapid (hasty) search techniques.

Since most SAR incidents are resolved by application of rapid (hasty) search tasks, likely routes of travel, attractions, hazards, etc., have high "historical" POC values. They also have very small areas, are generally free of obstructions, and it is possible to achieve very high POD values. Search speeds are high (often searchers literally "run" trails, for example) and effective searching can be done rapidly with few resources. It is after the rapid (hasty) search tasking comes up empty-handed that the full power of the optimal effort allocation method is required.

The techniques presented here are for when the decision is made to expand into area/grid searching in which allocation decisions become very involved and are no longer intuitive.

After several searches, the PSR of the segments should eventually all be approximately equal. Once the PSRs are equal, mathematically the optimal allocation is to spread the searchers throughout all search segments. The mathematics are correct that the maximum POS can be achieved in this manner, however that may not be operationally reasonable, safe, or smart.

Check the expected POS. By that point in the search, the expected gain in POS for any search is likely to be very small. Allocating searchers as larger teams in fewer segments is likely to generate a similarly small decrease in expected POS for the same or greater amount of effort put forth.

Consider cycling through groups of segments so that after a few search periods the PSR are all back to equal.

At this point in a search, the search planner could also begin to consider expanding the search area or suspending the search since there seems to be so little POC remaining that extreme effort is required to generate little increase in POS. A thorough re-evaluation of known information, clues, assumptions and scenarios explored is required to determine if another scenario is possible.

This page intentionally left blank.

Section 5-8: Probability of Detection (POD) Experiments

Overview

References

Actual Experiments

Overview

Effective Sweep Width (W) is the critical element to determining meaningful and accurate POD values. POD is essential for search planning and evaluating the effectiveness of a search task. This tool is then used to determine where to send future resources out into the field to maximize the POS. Simply put, knowing the Effective Sweep Width saves lives.

The only way to determine the Effective Sweep Width is to take searchers and perform field experiments in the specific environment(s) where actual searches take place.

There are three classes of factors that affect detection:

- The search object;
- The sensor in use; and
- The operating environment.

The purpose of a detection experiment is to determine the detectability of a particular type of search object in a particular environment when using a particular sensor.

The measure of detection for a given set of conditions (search object, sensor, and environment) is the Effective Sweep Width.. The Effective Sweep Width, together with information about the amount of effort that has been or may be expended in an environment, provides a means for objectively estimating the probability of detecting the search object if it is present. Effective Sweep Width provides a means for relating effort, area, and POD mathematically.

References

The recommended guide for conducting Effective Sweep Width experiments is available for download in three documents on the NSARC website (www.uscg.mil/nsarc):

- *A Simple Guide to Conducting Ground Search and Rescue Detection Experiments Volume I*;
- *A Simple Guide to Conducting Ground Search and Rescue Detection Experiments, Volume II, Appendixes*; and
- *Integrated Detection Experiment Assistant (IDEA)* software.

Actual Experiments

Conducting an Effective Sweep Width experiment requires planning, a large enough area, some limited equipment, searchers, data loggers, and the Microsoft

(MS) Excel® based Integrated Detection Experiment Assistant (IDEA) software.

The experiment can generally be conducted in one day. It will usually take from 1 to 4 hours for each searcher/data logger team to move through the experiment course. However, starting times may be more frequent so that several teams are on the track simultaneously, but not within sight or earshot of one another.

The following steps are required:

1. Design the experiment;
2. Find a search area;
3. Take environmental measurements;
4. Lay out the search track;
5. Place search objects;
6. Conduct the experiment;
7. Enter data; and
8. Analyze the results.

This guide along with the IDEA software will greatly assist in conducting a well-run and scientifically useful experiment. The experiment will benefit not only SAR planners but also searchers and ultimately the search object.

Almost all of the success of an Effective Sweep Width experiment rests with the planning and preparation. The IDEA software exists to help an experiment designer through all of the planning and data analysis stages. Perhaps the most important planning consideration is to keep the experiment as simple and short as possible. It is far better to have a successful experiment that determined a good Effective Sweep Width for one search object type than a long course with four different search object types under different conditions, but failing to provide any useful results.

Another important point to keep in mind is that in some respects, it is far easier to conduct an experiment than to describe one. Therefore, while these instructions may seem detailed and long, the experiment itself is actually rather straight forward. Table 5-8-1 provides the steps necessary for POD experiments.

Table 5-8-1: Instructions for Conducting POD Experiments

Pre-Experiment	Experiment Day	Post Experiment
1. Design the experiment; 2. Select candidate search areas; 3. Scout the search area; 4. Measure environmental factors; 5. Plan to lay out the search track; 6. Lay out the search track; 7. Plan to place search objects; and 8. Place search objects.	1. Establish the command center; 2. Perform the necessary logistics; 3. Instruct searchers and data loggers; 4. Dispatch the teams at appropriate intervals; and 5. Debrief the teams and collect data sheets.	1. Recover search objects; 2. Score results; 3. Enter data; 4. Determine Effective Sweep Width.

Section 5-9: Subsequent Searches

Cumulative Probability of Detection (POD_{cum})

Adjusting Probability of Containment (POC)

Calculating POD_{cum}

Cumulative Probability of Success (POS_{cum})

Suspend, Re-search, or Expand Search

Probable Success Rate (PSR)

Other Search Planning Considerations

Changing the Search Area

Moving Objects

Post-consensus Check

Research indicates the average missing person is found during the first operational period: about 8 - 12 hours (81% in 12 hours, 95% in 24 hours). However, when the search subject is not found, extended search planning may be required.

Cumulative Probability of Detection (POD_{cum})

A traditional method of tracking search progress is through calculating the accumulated effects on POD: cumulative POD, or POD_{cum}.

POD_{cum} in a search segment over many searches is found through its opposite: the cumulative probability the object would not have been found in any of the search periods $(1-POD)$. For any number of searches, the POD_{cum} can be calculated using the following formula:

$$POD_{cum} - [(1\text{-}POD_1)\ X\ (1\text{-}POD_2)\ X\ldots\ (1\text{-}POD_a)]$$

If POD_{cum} ever reaches 100%, the calculations are wrong. POD is calculated for individual segments and cannot be combined with POD of different segments. It is invalid to "average" the POD of several segments to show the "average" POD of the area. POD_{cum} is limited to showing how well an individual segment has been searched over time: the cumulative likelihood the searchers would have found the search object if it had been in the segment and able to be found.

Figure 5-9-1 below is an example calculation of POD_{cum}:

Example: POD_{cum}

POD of 1st Search: 35%

POD of 2nd Search: 30%

POD_{cum} is calculated as follows:

$$POD_{cum} = 1 - [(1-POD_1) \times (1-POD_2)]$$
$$= 1 - [1-0.35) \times (1-0.30)]$$
$$= 1 - 0.445$$
$$= 0.545 \text{ or about } 55\%$$

Figure 5-9-1: Example – POD_{cum}

Adjusting Probability of Containment (POC)

When a segment is searched and nothing is found, the search planner must revise the estimate of that segment's chances for containing the search object downward by an appropriate amount because some of the POC has been "searched out" of the segment.

After a segment has been searched, the likelihood the search object is still in the segment should be less than before it was searched. Less by an amount equal to the POS for the search in that segment. Figure 5-9-2 is the formula used to determine the POC remaining when POC is reduced by the POS of the search:

POC remaining is reduced by the POS of the search:

$$POC_{\text{remaining after the search}} = POC_{\text{before the search}} - POS_{\text{of the search}}$$

This formula is used to adjust POC after each search period.

To adjust POC only after several search periods, the equation becomes:

$$POC_{\text{after the series of searches}} = POC_{\text{before the series of searches}} - POS_{\text{of the series of searches}}$$

$$POC_{new} = POC_{original} - POS_{cum}$$

Figure 5-9-2: Formula for POC_{new}

It is important to use the same time frame for all terms in these equations. That is, the POS used must be that which is accumulated between the POC_{before} and POC_{after}.

Calculating POD_{cum}

POD_{cum} can be calculated using the formula in Figure 5-9-3.

$$POC_{new} = POC_{original} - (POC_{original} \times POD_{cum})$$

$$POC_{new} = POC_{original} \times (1 - POD_{cum})$$

$$POD_{cum} = 1 - \frac{POC_{new}}{POC_{original}}$$

Figure 5-9-3 Calculation of POD_{cum}

Figure 5-9-4 below is an example of how to calculate POD_{cum}:

Example: POD_{cum}

The following values apply:

$POC_{before} = 0.70$; $POD = 0.40$

Probability of Success (POS):

POS = POC X POD

POS = 0.40 X 0.70

POS = 0.28

Probability of Containment (POC) After the Search:

$POC_{after} = POC_{before} - POS$

$POC_{after} = 0.70 - 0.28 = 0.42$

Cumulative Probability of Detection (POD_{cum}):

$POD_{cum} = 1 - (POC_{new}/POC_{original})$

$POD_{cum} = 1 - (0.42/0.70)$

$POD_{cum} = 1 - 0.6$

$POD_{cum} = 0.40$

Figure 5-9-4: Example – POC_{cum}

Cumulative Probability of Success (POS_{cum})

POS_{cum} provides a measure of overall search effectiveness that considers the combination of both POD and POC.

POS is also additive across segments and across search periods, so it is much easier to calculate and track than other options.

For a segment, POS = POC x POD.

As shown below, in its simplest form, POS_{cum} is just the sum of the POS for the segments and/or for the search periods being considered.

$$POS_{cum} = POS_{cum\ of\ previous\ searches} + POS_{this\ search}$$

The basic equation can be modified to fit several specific circumstances, but the additive nature of POS remains.

As shown in the formula below, for a search segment (e.g., Segment 1) after several search periods, POS_{cum} is the sum of the individual period POS for the segment:

$$POS_{segment\ 1,\ cum} = \sum_{J=1}^{periods} POS_{segment\ 1,\ j} = POS_{segment\ 1,\ period1} + POS_{segment\ 1,\ period\ 2} + \ldots$$

For a region of many segments (or the entire search area) during a single period, POS is the sum of the individual segments POS and described in the following formula:

$$POS_{region} = \sum_{i=1}^{segments} POS_i = POS_{segment\ 1} + POS_{segment\ 2} + POS_{segment\ 3} + \ldots$$

It follows that the cumulative POS over several search periods is simply the sum of the POS of the individual regions for the search periods. This relationship is demonstrated in the following formula for POS_{cum}:

$$POS_{cum} = \sum_{j=A}^{regions} POS_{cum\ region\ j} = POS_{cum\ region\ A} + POS_{cum\ region\ B} + \ldots$$

POS_{cum} can never exceed 100% even with a 100% POD search; since POC is adjusted by POS after each period and the next POS is based on the POC remaining.

POS_{cum} will approach but never reach or exceed the original POC.

What does the math really tell the search planner?

- Doubling the effort by searching again does not double the likelihood of actually finding the search object;
- As the search continues, the same level of search effort produces less and less likelihood of finding the object in that period; and
- As the search continues, it takes increasingly larger amounts of search effort to produce significant increases in the likelihood of finding the search object.

Suspend, Re-search or Expand Search

POS_{cum} and POC remaining can be good indicators that either:

- The search object is not in the search area; or
- The search object cannot be found in the search area.

In either case, it reemphasizes the need for search planners and SMCs/ICs to periodically reevaluate the identified search area, search resources, and methods being used within the search area.

POS_{cum} and $POC_{remaining}$ may also be indicators of when to consider suspending a search, when the effort required becomes unreasonably high and is no longer in line with the expected benefits.

There is no rule for suspension, only indicators available for consideration.

Other Search Planning Considerations

Table 5-9-1 lists items that should be considered and may modify or change the search plan, regardless of the math:

Table 5-9-1: Other Search Planning Considerations

Objects: What is the object being searched for?	• Is the condition of the search object likely to have changed (e.g., responsive to unresponsive, mobile to immobile, etc.)? • Are the transient clues (e.g., smoke, scent, sign, and footprints after weather erosion, etc.) likely to be gone?
Urgency changes, alternative scenarios, and issues of survivability	• Are there other areas in which the search object may be located? • Are there areas that should be checked before survivability decreases too far? • If the search object were still moving, are there parts of the search area boundary he/she could have escaped through or new territory they could have moved into?
Clues	• Taken together, what do the clues indicate? • Do the clues change the search planning assumptions? • Include indications from any lack of clues in various areas.
SAR Facilities	Will re-searching a segment with a different type of SAR resource uncover more or different clues?

> Remember: Mathematical solutions are only guides and decision aids. The search planner should make decisions based on ALL available information.

Changing the Search Area

When searches become difficult or confusing, sometimes the best thing to do is start search area planning and searching over. Often, clues or information from either on or off scene sources can change the entire picture of the search, so much so that search planners/IC may want to re-evaluate where they believe the search object is located.

For example, in an aircraft search, conversion from an ASAD-based statistical search area into an information-based subjective-deductive search area may be required.

Sometimes, the team is just no longer comfortable with the defined search area and re-evaluates the clues available to decide if changes are required.

Revising consensus on POC may be a reasonable method to update the search planning effort. As necessary:

- Discuss alternative scenarios, particularly from the perspective of how the new information challenges or changes the assumptions within each of the scenarios;
- Carefully redraw the search area and regions; and
- Use a consensus process to establish a current set of POC values and update the probability map and search plan.

The following are two procedures that may be used for revising consensus.

1. *Take the results of all previous no-find searches into account.*

With this option, consensus is revised with consideration of all of the segments already searched. The consensus participants use that knowledge to establish where they believe the missing person is currently located and the associated POC.

It seems reasonable to assume the participants will intuitively adjust their POC assignments based on that previous searching. If a current scenario would lead to a region containing 50% POC, but has been extensively searched already, the search planners may assign significantly less than 50% POC to that region.

Since the POC has been adjusted intuitively, no further mathematical adjustments are necessary. The consensus POC values represent the best judgment of the leadership team, taking into account all relevant factors.

2. *Do <u>not</u> take the results of any previous searches into account.*

With this option, the consensus is revised consciously ignoring the fact that some segments and regions have already been searched. The effect of previous searches is taken into account afterwards.

Consensus participants make their judgments based only on the information and clues currently available. Once consensus is reached, POC can be adjusted using the equations already described in this chapter.

For each segment:

$$POC_{corrected} = POC_{from\ consensus} \times (1 - POD_{cumm\ effects\ of\ previous\ searches})$$

If using the second procedure to accomplish consensus, care must be taken to ensure consensus participants can actually distance themselves from the knowledge of previous search efforts sufficiently to arrive at honest POC judgments in this manner. If they cannot, the second method would over-correct POC by double counting the effects of previous searches: once intuitively, then again mathematically.

Moving Objects

If the search object is moving, search planners should start over each period and re-establish where they think the object will be during the next period (POC).

With a moving object, the effects of previous searches have no impact on the current POC. The revised consensus must ignore the fact that some segments and regions have already been searched.

The resulting consensus POC is not "corrected" for previous POD.

In addition, POD_{cum}, POS_{cum}, and $POC_{adjusted}$ are not valid in a moving object case.

POD, POS, PSR, and POC only have meaning for individual search periods with moving objects.

Post-Consensus Check

In any case, it is wise to check the consensus results for reasonableness before continuing with search planning. As with the original consensus, ensure:

- Each participant is comfortable with the results;
- The results reflect a reasonable composite of the expert judgments involved; and
- The results pass the collective common-sense test.

A faulty set of POC values will adversely affect the entire search.

Appendices

This page intentionally left blank.

Appendix A: Comparison - International SAR System vs. NIMS/ICS

Table A-1 below provides a comparison between NIMS/ICS and the international SAR system in use in the U.S. maritime and aeronautical SRRs.

Table A-1: Comparison – International SAR System vs. NIMS/ICS

	Characteristic	International SAR SYSTEM	NIMS/ICS
1.	Users	SCs, SMCs, and SAR facilities	SAR responders and NIMS/ICS incident management organizations
2.	Applicability	Worldwide for aeronautical and maritime SAR, and within the U.S. for all types of civil SAR covered by the NSP. *(Note: The U.S. civil SAR system is essentially an expansion of the global SAR system of which it is an integral part; a global SAR system avoids the necessity for each country to maintain SAR services for its own citizens wherever they travel.)*	Mainly within the U.S. at all government levels.
3.	Primary legal framework	Chicago Convention, International Convention on Maritime SAR, various international agreements to which the U.S. is Party, and the NSP.	Stafford Act, HSPD-5, and the NRF with its Annexes.
4.	Who needs working knowledge	Mainly Federal (including military) SAR personnel	SAR responders and incident managers at the Federal (including military), State, Tribal, Territorial, and local levels.
5.	Primary implementing guidance	IAMSAR Manual and the NSS and its associated Addenda	Department of Homeland Security NIMS Document and the NSARC's CISAR Addendum.
6.	Operational applicability	All types of civil SAR, including CISAR	Full spectrum of emergency incidents and hazardous situations
7.	What is covered	Concepts and principles, terminology, assignment of responsibilities, organizational structure, operational policies, preparedness, resource management, operational coordination for single and multiple organizations, information management, training, exercises, and agreements	Essentially the same as the international SAR system.
8.	Geographic responsibilities	Internationally-recognized SRRs	Political jurisdictions
9.	Terminology	Per the IAMSAR Manual and NSS glossaries and acronyms	NRF Resource Center glossary and acronyms; much of this terminology is associated with ICS

Table A-1: Comparison – International SAR System vs. NIMS/ICS (continued)			
	Characteristic	**International SAR SYSTEM**	**NIMS/ICS**
10.	**Size of incidents covered**	All scales of SAR operations.	All scales of SAR operations.
11.	**Main operational response structures**	RCCs and RSCs, with an SMC assigned to the case and an OSC at the scene of distress	Incident, unified and area commands, usually near the incident scene, with an SMC/IC assigned
12.	**Use of incident command system (ICS)**	In most cases involving State and local authorities, in all cases involving CISAR, and as required for larger cases (particularly for mass rescue operations (MROs) involving multiple U.S. agencies.	Essentially all cases where suitable.
13.	**Organization structure**	Typically centered on Federal RCC coordination, but with delegation to lower agency levels, or to States in accordance with Federal-State agreements. There is one RCC for each SRR.	Response coordination is typically done at the local level if the capability exists there, with support from higher government levels, with Incident, Unified or Area Commands set up in accordance with ICS.
14.	**Preparedness**	Planning, training, exercise, personnel qualification and certification, resource acquisition, agreements, and publication management	Same as for the international SAR system, except involves more equipment certification and typing of SAR resources.
15.	**Resource management**	Personnel and equipment are provided mainly to meet requirements based on risk assessments, and are used, directed, tracked, and replenished over the duration of a SAR case (mission or incident)	Processes covered to describe, inventory, mobilize, track, and recover resources
16.	**Agreements**	SAR authorities are supported by a wide array of national agreements and contracts mostly at the Federal level, including the NSP, but also at lower levels, and by numerous multi-lateral and bilateral international agreements, including conventions (treaties)	All States, the District of Columbia, Puerto Rico, Guam and the U.S. Virgin Islands are voluntary signatories to EMAC to support mutual assistance; except for a few international agreements, other services provided in relation to NIMS consist of an array of agreements and contracts at all government levels.
17.	**Financing sources**	Public taxes are the main funding source for civil SAR; however, substantial commercial fee-for-services and corporate or volunteer services-in-kind resources are provided both nationally and internationally	Most NIMS services are funded by public taxes; however, volunteer and corporate contributions have offset some costs, especially in large-scale incidents.

Table A-1: Comparison – International SAR System vs. NIMS/ICS (continued)			
	Characteristic	**International SAR SYSTEM**	**NIMS/ICS**
18.	**Agreements**	SAR authorities are supported by a wide array of national agreements and contracts mostly at the Federal level, including the NSP, but also at lower levels, and by numerous multi-lateral and bilateral international agreements, including conventions (treaties)	All States, the District of Columbia, Puerto Rico, Guam and the U.S. Virgin Islands are voluntary signatories to EMAC to support mutual assistance; except for a few international agreements, other services provided in relation to NIMS consist of an array of agreements and contracts at all government levels.
19.	**Financing sources**	Public taxes are the main funding source for civil SAR; however, substantial commercial fee-for-services and corporate or volunteer services-in-kind resources are provided both nationally and internationally	Most NIMS services are funded by public taxes; however, volunteer and corporate contributions have offset some costs, especially in large-scale incidents.
20.	**Documentation**	A blend of international and national publications, plans, agreements, and forms sponsored by many different nations and organizations.	Mainly developed, named, promulgated and managed at all government levels in accordance with NIMS/ICS
21.	**Communication**	Distress and SAR coordination communications, including equipment performance standards, are mostly based on regulations and standards at both international and national levels to address standardized frequencies, equipment performance and registration, and message formats.	Similar to the international SAR system, except that most regulatory provisions are national, and need for interoperability and inter-linking are major considerations.
22.	**Information management**	Except for MROs and CISAR, information management depends on typical office capabilities, the internet and specialized software. NIMS requirements must be taken into account.	NIMS information management is usually routine, but must be characterized by robust capacities that function even when local technologies and links are knocked out by an incident, that accommodate massive amounts of information from many sources, and that can present the right information at the right time to the right persons even with information loads at or near saturation levels for the systems used.
23.	**Operational coordination**	The person with overall responsibility for coordinating a SAR response is normally the SMC who works in an RCC; however the SMC function can be performed at a level lower than the RCC if the situation does not require RCC authorities and capabilities, or to a higher level than the RCC if it is necessary to raise the situation.	The person with overall responsibility for coordinating an incident response under NIMS is an IC, who is often a local authority operating in an incident, unified or area command, depending on which is established to deal with the incident.
24.	**Scalability**	The international SAR System is designed to accommodate all types of scenarios that might occur in a particular SRR, and an RCC can handle most scenarios.	NIMS routinely depends on an ICS structure that is highly scalable. Scalability might include establishing and disestablishing unified and area commands.

Table A-1: Comparison – International SAR System vs. NIMS/ICS (continued)			
	Characteristic	**International SAR SYSTEM**	**NIMS**
25.	**Compatibility of the international SAR system and NIMS/ICS**	At the Federal level, because of the use of fully staffed 24-hour RCCs, the international SAR system normally does not use NIMS. However, Federal SAR authorities will commonly use NIMS when working with non-SAR authorities or entities regardless of whether or not CISAR is involved, and with local, State or Tribal authorities. The NPS uses NIMS for nearly all of its SAR operations. SAR personnel are required to have a working knowledge of NIMS.	All land-based SAR operations within the continental U.S. and Alaska normally use NIMS/ICS. For CISAR under the Stafford Act/NRF, an SMC is normally assigned and "plugs" into NIMS/ICS in the Operations Section. It is the SMC (normally the SMC is the State) or emergency manager that will coordinate the SAR response under the NIMS/ICS structure.

Appendix B: Cospas-Sarsat Doppler Location Technique

Doppler Location

Frequency Difference of Arrival (FDOA) and Time Difference of Arrival

Doppler Location

The Cospas-Sarsat low Earth orbit (LEO) satellite system determines the location of a distress beacon using Doppler frequency shift between the active distress beacon and an orbiting satellite (based on two principles of geometry illustrated in Figures B-1 and B-2).

Figure B-1 shows the frequency of the distress beacon as received by the satellite vs. time for two satellite passes (1 & 2).

Pass (1) depicts the maximum elevation angle of the satellite (as viewed from the distress beacon) about 60° above the local horizon. At time t_0 (when the satellite first comes into view on the horizon) the distance between the distress beacon and the satellite is at its maximum. At this time, the velocity of the satellite towards the beacon is at its maximum, resulting in an apparent increase of the distress beacon carrier frequency (termed Doppler shift) as measured at the satellite. As the satellite elevation angle increases, its velocity relative to the distress beacon decreases, resulting in a decreasing Doppler shift as measured at the satellite.

This process continues until time t_{tca} (TCA: Time of Closest Approach), when the satellite is at its maximum elevation angle as viewed from the distress beacon. This is the TCA between the satellite and the distress beacon. At this moment the velocity vector from the satellite towards the distress beacon passes through zero, and becomes negative, and the satellite begins moving away from the disress beacon, resulting in a change of sign of the Doppler curve (the inflection point).

As the satellite continues its orbit, the velocity vector away from the distress beacon continues to increase, becoming greatest as the satellite disappears below the local horizon (when the Doppler shift is at its negative maximum).

Pass (2) depicts a low elevation pass where the maximum elevation angle of the satellite is approximately 20 degrees above the local horizon.

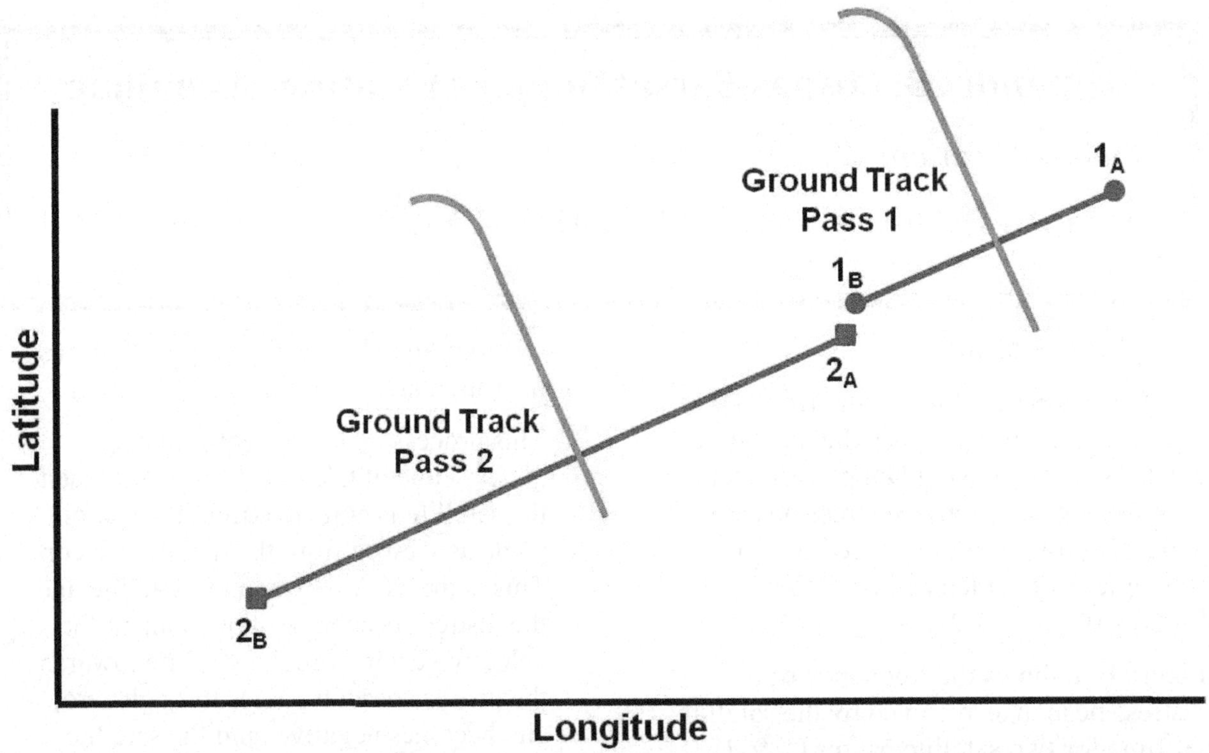

Figure B-1: Distress Beacon Frequency Received by Satellite vs. Time for Two Satellite Passes

Figure B-2 shows the ground tracks of two consecutive passes (1 and 2) of the same satellite, which occur about 100 minutes apart. Since the orbit of the satellite is precisely known, the point of TCA along the ground track can be easily determined. Also, the slope of the Doppler curve at TCA allows the distance and direction (either east or west) that the distress beacon is from the satellite ground track at TCA to be determined. This provides two possible solutions for the distress beacon location, 1_A and 1_B. A second satellite pass provides another set of possible solutions, 2_A and 2_B, which resolves the ambiguity of a single satellite pass, in this example, the location indicated by 1_B and 2_A.

An additional component of Doppler shift is the effect of the rotation of the earth, which causes an increase in Doppler shift if the distress beacon is West of the satellite ground track, and a corresponding decrease when the distress beacon is east of the satellite ground track. This effect is greatest if the distress beacon is located at the Earth's Equator, and decreases to zero if the beacon is located at either Pole. The Ground Station Doppler processor accounts for this small Doppler shift and uses it to calculate the probability that the "A" solution is the True location (and conversely that the "B" solution is actually an image). The ability to make this determination depends on the beacon signal being very stable (as is the case with Cospas – Sarsat 406 MHz distress beacons), and on the distress beacon location being in low to medium latitudes.

Experience has shown that with 406 MHz distress beacons, the side "A" probabilities generally exceed 70%, and often are above 90%, allowing search and rescue resources to be committed even before the A--B ambiguity is resolved by a second satellite pass.

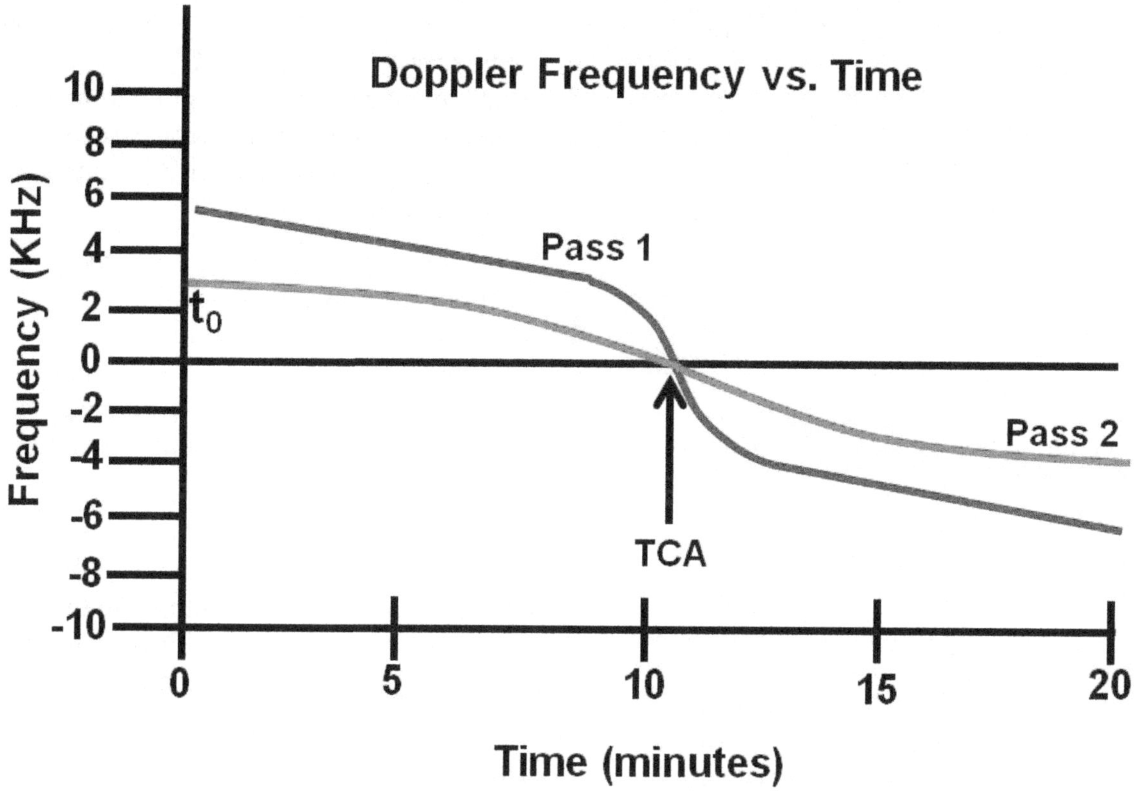

Figure B-2 Doppler Plot

Frequency Distance of Arrival (FDOA) and Time Distance of Arrival (TDOA)

The future of the International Cospas-Sarsat System is the ongoing development of the medium Earth orbit (MEO) SAR satellite system (in the U.S. this is known as the Distress Alerting Satellite System – DASS), which uses Frequency Difference of Arrival (FDOA) and Time Difference of Arrival (TDOA) techniques to locate an active distress beacon.

FDOA, also frequently called differential Doppler (DD), is a technique analogous to TDOA discussed below for estimating the location of a radio emitter based on observations from other points. MEOSAR uses TDOA and FDOA together to improve location accuracy. By combining TDOA and FDOA measurements, instantaneous geo-location can be performed in two dimensions.

FDOA differs from TDOA in that the FDOA observation points (orbiting satellite receivers) must be in relative motion with respect to each other and the distress beacon. This relative motion results in different Doppler shift observations of the distress beacon signal. The distress beacon location can be estimated with knowledge of the satellites' exact locations and vector velocities and the observed relative Doppler shifts between pairs of locations.

A disadvantage of FDOA is that large amounts of data must be moved between the satellites and a single MEO Local User Terminal (MEOLUT) to do the cross-correlation needed to estimate the Doppler shift.

The accuracy of the location estimate is related to the:

- Bandwidth of the distress beacon signal;

- Signal-to-noise ratio at each satellite receiver; and the
- Geometry and vector velocities of the distress beacon and the satellites.

Multilateration, also known as hyperbolic positioning, is a process that MEOSAR will use to locate a distress beacon by accurately computing the TDOA of a signal emitted from the distress beacon to three or more satellite receivers.

A pulse emitted from a distress beacon will arrive at slightly different times at two spatially separated satellite receivers, the TDOA being due to the different distances of each receiver from the beacon. With two receivers at known locations, an emitter can be located onto a hyperboloid. The receivers do not need to know the absolute time at which the pulse was transmitted - only the time difference is needed. (A Hyperboloid is quadric surface generated by rotating a hyperbola around its main axis, and a hyperbola is an open curve formed by a plane that cuts the base of a right circular cone.)

A third receiver (satellite) at a third location would provide a second TDOA measurement and locate the emitter on a second hyperboloid. The intersection of these two hyperboloids describes a curve on which the beacon is located.

With a fourth receiver, a third TDOA measurement is available and the intersection of the resulting third hyperboloid with the curve found with the other three receivers defines a unique point in space. The beacon's location is therefore determined in 3D.

In practice, errors in the measurement of the time of arrival of pulses mean that enhanced accuracy can be obtained with more than four receivers. And, the TDOA of multiple transmitted pulses from the emitter can be averaged to improve accuracy.

Encoded Locations

Some distress beacons can acquire and relay their own positions using the Global Positioning System (GPS) or some other global navigation satellite system (GNSS). The locations are encoded in the beacon signal and forwarded with the distress alert.

Appendix C: SAR Pre-Plan Elements

I. Environment

II. Resources

III. First Notification Procedures

IV. Action Plan Development, Guides, and Procedures

V. Support Requirements Plans Development

VI. Post Mission

Note: What follows is a sample list of SAR pre-plan elements. These elements can serve as a starting point for organizations developing unique SAR pre-plans. This Appendix is not intended to be a complete listing of all information requirements or sources. Organizations must tailor their SAR pre-plans to their own missions, environments, and requirements.

I. Environment

- Maps (in a variety of scales) highlighting:
    - Hazards and danger areas;
    - Attractions where visitors tend to explore;
    - Roads, trails, facilities; and
    - Locations of past incidents.
- Descriptions of:
    - SAR vulnerabilities: how people have found troubles before;
    - Risk analysis: other potential trouble areas/events;
    - Area considerations:
        - Weather and seasonal changes that might affect SAR and objectives; and
        - Potential hazards to SAR resources –terrain, vegetation, wildlife, etc.

II. Resources

- Who/what SAR resources are available;
- Specialties: specifically what they can provide, response options available to you;
- Qualifications, certifications and training;
- Limitations (capabilities, availability, special needs, reimbursement, legal restrictions);
- Expected response times;
- Contacts:
    - Contact for permission to use resource;
    - Current list of who and how to contact;
    - List of "initial info" requirements; and
    - What the resource requires.
- Human and animal resources examples:
    - Searchers;
    - Dogs teams; and

- Specialties (man tracking, water, caves, technical climbing, air, electronic).

- Equipment resource examples;
- Vehicles, water craft, aircraft (ex. hoist);
- SONAR, FLIR;
- Tractors, cranes, other heavy equipment;
- Informational resource examples (incl. own agency);
- Investigative: RCC, FAA, media, law;
- Data resources:
    - AOPA - airport listings;
    - Media;
    - People with local/historical knowledge; and
    - Local campground/attraction contacts.
- Support resource examples:
    - Local law enforcement;
    - State SAR coordinating agencies;
    - Local SAR support specialists (plans, logistics, communications);
    - Communications (phone, HAM radio, etc.);
    - Food, water, shelter for searchers: Red Cross, Salvation Army; and
    - Medical: for survivors and for searchers.

III. First Notification Procedures

- Forms for collecting initial information;
- Guidelines for taking the initial call:
    - Questions to ask; and
    - Attitude and how to ask necessary probing questions.
- Guidelines for initial evaluation:
    - Urgency;
    - Who is in charge?
    - Initial search area, method, resources; and
    - Investigation (outside of search area).

IV. Action Plan Development, Guides and Procedures

- Documentation and record-keeping: data, actions, decisions, judgments;
- Continued investigation/collection of data;
- Comparing current with historical incidents;
- Initial callout;
- Establishing SAR organization;
- Establishing and managing base camp;
- Managing search teams:
    - Arrival;
    - Assignments and briefings; and
    - Debriefings, did they return?
- Planning search strategy and monitoring progress:
    - Initial containment;
    - Initial plan: areas and application of appropriate resources (based on informed POC and POD & POS); and
    - Revising plan and expanding search area.
- Rescue/recovery (stabilize/transport); fatalities;
- Guidelines for suspension;
- Demobilization plan;
- Transfer of control (scope, jurisdiction).

V. Support Requirements Plans Development

Who, what, where, when, how?

- Logistics: Food, water, shelter, facilities, power, heat, sanitation;
- Communications: base-world, base-search teams, team-team;
- Medical: for victims and for searchers;
- External influences: media, family; and
- MOA/MOU with other agencies.

VI. Post Mission

- Critiques and lessons learned;
- Documentation;
- Review and update of plans, procedures, and agreements; and
- Preventive education.

This page intentionally left blank.

Appendix D: Search Urgency Determination Form

Table D-1 on the following page is the Search Urgency Determination Form. This form provides an analysis to help search planners decide the urgency of searching for a particular person. The lower the rating correspond to higher the urgency. The numerical range of urgency is from 7 to 21.

Consider any category scored a "1" as potential high urgency.

Use the below scale to score the search urgency.

7	9	11	13	15	17	19	21
Highest Urgency			Intermediate Urgency				Lowest Urgency

Table D-1: Search Urgency Determination Form		
Subject and Profile	**Rating**	**Score**
Age:		
Very Young	1	
Very Old	1	
Other	2-3	
Medical Condition:		
Known or suspected injured or ill or mental problem	1-2	
Healthy	3	
Known fatality	3	
Number of Subjects:		
One	1	
More than one (unless separation suspected)	2-3	
Subject Experience Profile:		
Not experienced, does not know area	1	
Not experienced, knows area	1-2	
Experienced, does not know area	2	
Experienced, knows area	3	
Weather Profile:		
Past and/or present hazardous weather	1	
Predicted hazardous weather (8 hrs. or less)	1-2	
Predicted hazardous weather (more than 8 hrs.)	2	
No hazardous weather predicted	3	
Equipment Profile:		
Inadequate for environment and weather	1	
Questionable for environment and weather	1-2	
Adequate for environment and weather	3	
Terrain/Hazards Profile:		
Known hazardous terrain or other hazards	1	
Few or no hazards	2-3	
Total Score		
Your Name:	**Date/Time:**	

Appendix E: Lost Person Questionnaire

Table E-1 on the following pages is a sample Lost Person Questionnaire. This questionnaire provides a format to collect information in an efficient and coherent manner and is used to support a missing person investigation. It may also be used as a guide for conducting an interview.

Information used to complete the form may come from multiple sources and interviews. Use a separate form for each interview and collate the information in a master file. If there is more than one missing person associated with this case, use a separate form for each person.

Table E-1: Lost Person Questionnaire

Case Name/Number			Agency
Date	Time	Location	
Interviewer's Name		Title	Agency
Information given by			DOB
Address			
Home Phone		Business Phone	
Cell Phone, Other Numbers			
Occupation		Employer	
Relationship to missing person			
Other persons interviewed: Name, contact information, date, time & relationship.			

Missing Person

Full Name				Nickname(s)
Name to call				Aliases
Safe word?	Y	N	Word	Who knows it?
Subject's primary language				
Home address				
Business or local address				
Home Phone				Business Phone
Cell Phone, Other Numbers				
E-mail address				

Description

Age	Race	Gender	Ht	Wt	DOB
Build					

Hair Color	Length	Style
If balding, describe		
Describe facial hair		

Eye color	Glasses	Y	N	Regular	Sun	Contacts
Describe glasses						
Eyesight without glasses						
Facial features, shape						
Complexion						
Distinguishing marks, tattoos, scars						
General appearance						

Clothing Worn When Last Seen
Hat/Cap/Scarf
Shirt/Blouse
Pants
Dress
Sweater
Coat/Jacket/Raingear
Footwear
Hose/Socks
Underwear
Glasses
Other
Describe all accessories the subject may have been wearing, such as belt, rings, watch, pins, hair accessories, necktie, tie clip, etc.
Describe all items the subject may have been carrying, such as pocketbook, wallet, backpack (describe contents of each), cell phone, keys, pocket knife, pager, camera, weapon, etc.

Details of Loss	
Location missing from	
Point Last Seen (PLS)	
Day/Date Last Seen	Time Last Seen
Last seen by whom	
Last communication (Date/Time/Method)	
Subject accompanied by animal(s)? Describe	
Vehicle description, if driving	
Destination(s), stated intentions	
Possible routes	
Weather at time of loss	
Events of last 24 hours leading up to time of loss	
Reported missing by:	Why?
Address	
Phones	
Relationship to missing person	
Where can this person be reached in the next 12 hours?	

Subject's Experience

Resident of	How long?
Previous residence	How long?
Birthplace	
Has this person been the object of a search in the past?	
If so, describe date(s), circumstances of loss, how long missing, when found, where found, condition when found and actions taken by subject while missing (if known)	
Additional Information and Comments	

Physical Health

General physical condition		
Handicaps		
Known medical problems		
Pregnant?	How long?	Menstruating?
Physician		Phone
Address		

Mental/Emotional Health	
General mental health	
Known mental problems	
Suicidal?	Previous attempts (explain)
Is this subject possibly dangerous to self or others? Explain:	
Does this subject have access to or is he/she possibly carrying a weapon?	
Are all weapons accounted for?	
Fears and phobias:	
Knowledgeable person	Phone
Address	

Medications: Prescription and Non-prescription	
Medication, strength and dosage	Affect if not taken

Identification

Drivers License:	State	No.	Date Issued
Other Identification			

Is subject enrolled in Safe Return or similar program?		Describe:
Electronic tracking device?	Describe:	

Finances

Credit cards: List card names and account numbers	

Checking and savings accounts: List banks and account numbers			
Does subject have credit cards or check book in possession?	Y	N	Cash carried:
Describe:			

Detailed Subject History

Single		Married		Divorced		Widowed

Spouse's Name	Phone

Address (if different)

Siblings (Name, age, residence) Use Back If Necessary

Father's Name	Living?	Y	N

Contact Information

Occupation & Employer

Mother's Name	Living?	Y	N

Contact Information

Occupation & Employer

Other relatives that may provide information

Subject's primary occupation	Retired?	Y	N

Employer	How long?

Contact person

Previous employment history

Education level

Detailed Subject History (continued)

Military service branch	Currently active or reserve?	Y	N
Contact person	Dates of service		
Religion or belief system	Active?	Y	N
Contact person			
Other persons who may provide information			
Hobbies, special interests			
Experience in outdoors, backcountry			
Favorite places to visit			
Athletic ability, mobility			
Active/outgoing or quiet/withdrawn?			
Attitude toward authority			
Recent, current or anticipated financial, legal or other problems			
Who does subject confide in and/or whom does he/she frequently talk to on the phone?			
Who last talked with subject at length?			
When and what was topic?			
Recent letters or writings?			

Detailed Subject History (continued)

Does subject keep a diary?	
Does subject have access to a computer?	Describe locations, user name(s), password(s)
Does subject smoke, drink or use illegal drugs? Describe in detail:	
Additional Information and Comments	

Children, Elderly, Special Needs				
Refer to mental/emotional health section			Mental age, if known	
How old does the subject look?				
Fears and phobias:	Horses?	Dogs?	Dark?	Sirens, loud noises?
Other: (describe)				
Will subject answer, if called?		Preferred name to call		
Any training on what to do if lost, such as Hug-A-Tree?				
How does subject normally travel? (Foot, bike, public transportation, family, friends, etc)				
Will subject talk to strangers, accept rides?				
Is there a "home place" or other special place?				
Does subject have a caretaker or a day care facility?				
Can the subject dress and/or feed him/herself?				
Does there appear to be any issues with family, school or care facility?				
Does the subject know how to call home or call 911?				
What would this subject most likely do if lost?				

Additional Information and Comments

Appendix F: Initial Missing Person Checklist

(Reference: National Park Service Field Search and Rescue Manual)

The Initial Missing Person Checklist is a general checklist for responding to a missing person report, and may have already been started or completed by the time SAR resources arrive at an incident. This is a general checklist and not all inclusive, some items may not apply or additional steps may be required. The order of actions may need to be changed to meet the needs of a particular incident. What is imperative is for search efforts to begin without delay.

To clarify: dispatch search resources to any report of the subject's LKP/PLS, or stated intentions indicate a woodland or wilderness search area in which the subject, due to loss or mishap, may be unable to return to safety on their own.

- Dispatch Actions:
  - Determine a call-back number and name of the reporting party. Also quickly identify address/location and relationship to missing subject.
  - Use the Lost Person Initial Questionnaire (*Appendix E*) as a guide to collect additional information.
- Initial Report Collection:
  - Complete the Search Urgency Determination Form, if desired (*Appendix D*).
  - Dispatch an appropriate resource.
- Initial actions upon being assigned by a dispatcher:
  - Collect a brief initial report from dispatch. Information should include:
    - Age, sex, physical and mental condition of missing person(s);
    - LKP/PLS, time last seen, circumstances of loss, subject activity; and
    - Adequate clothing, brief description.
  - Obtain weather information.
  - Conduct or consider Risk Assessment for conducting operations.
  - Obtain directions to the incident and check against a detailed map if any questions.
  - Determine types of resources that might be available to assist.
  - Review initial information and determine if requesting additional SAR resources is required prior to arrival on-scene.
  - Request appropriate equipment be dispatched to the incident (e.g., medical equipment, rescue equipment, search management equipment, logistical support).
  - Requested resources are alerted, if appropriate.
  - Inform the dispatching agency of Incident Commander estimated time of departure, when you actually leave, and estimated time of arrival to incident site.
- While en-route to the incident:
  - Receive updated information from dispatch or coordination center prior to arriving at the incident.

- Receive or request updated information on availability of special resources (e.g., rescue teams, dog teams, or aero medical helicopters).
- Obtain latest weather forecast if not already obtained.
- Inform dispatching agency when arriving on-scene.

• Upon arrival. The order of actions to take upon arrival will change depending upon the unique circumstances of the search. The Incident Commander must be highly flexible during the first hours of an incident. The following list serves merely as a suggestion for initial tasks. The order of these steps will vary from incident to incident.

  - Meet or talk with reporting party.
  - Review initial information, collect pertinent portions of the Lost Person Questionnaire *(see appendix F)*.
  - Consider the probability of the subject being lost, abducted (Amber/Silver Alert), runaway, elder abused, committed suicide, criminal event, etc.
  - Determine if a physical search is warranted;
  - Implement ICS.
  - As the situation warrants, consider transfer of command.
  - Revise urgency of search based upon any new information.
  - Determine initial resource requirements.
  - Determine Incident Command Post (ICP) location and staging areas. Consider the following requirements for selection:
    - Electricity and lights (generator may be required).
    - Work area sheltered from weather, media, and family.
    - Telephone or area with solid cell phone coverage.
    - Copier/Scanner.
    - Radio communications.
    - Running water if large or multi-operational period search.
    - Sanitation if large or multi-operational period search.
    - Staging area.
    - Parking area.
    - Heli-spot, if required.
  - Request or alert resources including specialized resources.
  - Appoint Operations Section Chief (OSC) and Investigators from the most qualified individuals present. Activate other ICS elements, as required.
  - Investigations:
    - Obtain photograph of the search subject and prepare information flyer. Once information flyer is prepared, review with IC/PIO.
    - Complete Lost Person Questionnaire (*Appendix F*).
    - Issue radio report to surrounding jurisdictions, if appropriate.
    - Contact family and friends of search subject.
    - Contact local hospitals.
    - Contact local transportation hubs.
    - Contact Emergency Services Agencies.
    - Contact local shelters and jails.
    - Contact morgues and medical examiners' office.

- Consult Lost Person Behavior Statistics/Information.
- Develop initial search objectives and strategies.
- Ensure tentative medical/evacuation plan developed prior to deploying resources.

- Plan and Deploy Initial Actions.
  - Conduct an initial briefing.
  - Quickly review subject information, possible scenarios, and any relevant clues.
  - On a detailed map mark the Initial Planning Point (IPP). This may represent either the LKP or PLS.
  - Quickly determine primary search area based on theoretical, statistical, subjective and deductive information (See Chapter 5 - Search Planning for more information).
  - Perform a quick consensus to determine priority areas.
  - Mark LKP/PLS and/or IPP on the planning map. Commence search.
  - Mark confinement tasks on the planning map.
  - Mark rapid (reflex/hasty) search tasks on the planning map.
  - Mark high probability areas on the planning map.
  - Deploy Resources.
- Continued Initial Search Efforts and Management:
  - Coordinate staff activity.
  - Approve the use of different training levels on the incident if required.
  - Ensure the health and safety of resources already deployed. Monitor fatigue levels.
  - Ensure efficient flow of personnel from staging to field.
  - Identify and meet with liaisons/agency representatives to keep informed and determine any special capabilities and requirements.
  - Keep agency administrator and coordination center informed of resource requests, current situation, and information updates.
  - Constantly be available to general staff.
  - Work to create an environment for staff to work in. Shield staff from family, media, and political pressures.
  - Drive mission, identify problem areas, influence overall direction of mission, establish initial objectives, and enforce priorities.
  - Verify that everyone on general staff is being kept up-to-date.
  - Verify that information is flowing within command structure.
  - Ensure clues and investigative information is being documented and followed-up.
  - Determine when to start next operational period and begin extended operations.

Remember: the Initial Missing Person Checklist is not all inclusive; each search is different and will have unique characteristics. Use of the checklist will assist personnel and agencies in getting a search started, but responsible parties will need to make additions or deletions to meet the needs of a specific incident.

This page intentionally left blank.

Appendix G: Lost Person Behavior Category: Dementia (Alzheimer's)

Introduction

Dementia includes Alzheimer's disease and several other related disorders (vascular dementia, Parkinson's disease, and dementia with Lewy bodies among others). From a search and rescue perspective there is no appreciable difference between the different dementias. Alzheimer's disease is the most common form of the irreversible dementias and the term is included since, in common language, it is often used to represent all dementias. Regardless of activity, search subjects with dementia were placed into this category.

Dementia results in a wide range of changes in behavior and cognitive (thinking) skills. It ranges from mild, where the person is still able to perform all activities of daily living, to severe. The more severe the dementia, the more likely the subject is to wander and experience symptoms such as hallucinations and psychosis. Dementia is the loss of memory, reason, judgment, and language to such an extent that it interferes with daily living. Some of the earliest symptoms may be problems with short-term memory. One or more cognitive areas are disturbed: (1) aphasia (problems with language, e.g., finding the right word); (2) apraxia (cannot move body correctly); (3) agnosia (cannot recognize common objects, especially faces); (4) decreased executive functions (planning, organizing, abstracting). Dementia often results in severe disturbances in how a person perceives and interprets events, sights, and sounds around him.

Furthermore, the visual field (peripheral vision) is narrowed, creating "tunnel vision." A reduction in peripheral vision results in poor navigators using only what they see in front of them versus excellent navigators who combine active scanning for landmarks they remember. This may account for dementia wanderers' trademark behavior of essentially moving straight ahead: "They go until they get stuck." Direction of travel predicts a dementia subject's final location better than it does in most other subject categories.

Every subject is different, and determining the severity of the individual's dementia provides additional important insight. Mild to moderate severity is associated with more goal-directed wandering. Initially, the subject does in fact have a destination in mind. Distances traveled may be greater. The subject is more likely to use public transportation. In a short conversation with the subject one may not detect anything unusual. Dementia subjects are usually recognized by the public due to inappropriate dress, unsafe or inappropriate behavior, asking for assistance, or an inappropriate response. If engaged in a conversation, then suspicion is usually raised by repetitive questions, phrases or words. Subjects with severe dementia tend to show random (i.e., no discernable goal) wandering, travel shorter distances, and may have profound sensory disturbances. Exit-seeking behavior may be seen in both mild and severe dementia. One is more likely to see this behavior when a person is in a new location or has been taken out of a familiar environment.

Dementia Hallmark Behaviors

1. They go until they get "stuck."
2. Appear to lack the ability to turn around and may Ping-Pong off some barriers.
3. Direction of travel is a good predictor of where they are found.
4. Look for sign and attempt to determine exit door.
5. Oriented to the past. The more severe the dementia, the further in the past they exist. Figure out where in the past the subject is currently "living" in order to determine possible destinations (e.g., a former residence, a work place). Investigative questions assist to better understand the subject's past (which, for them, may be the present).
6. May attempt to travel to former residence, favorite place or what appears to be former place, or workplace.

In an urban environment, the subject is typically found in structures or walking along roads. In both urban and wilderness environments, the subject is highly likely to cross or depart from a road (66%). If the subject leaves the road or travel feature, he does not travel far. He may often go unnoticed unless his dress is highly unusual. Track offset statistics for dementia are the shortest of all subject categories. In the wilderness, the subject is typically walking or gets stuck in brush/briars or drainages. Structures are common. Subjects are also attracted to water features and will walk into water (perhaps without even realizing it is water). They are generally mobile for only a short period of time. In temperate domains, half of the subjects are mobile for less than an hour. In dry domains, subjects remain mobile longer. They will not leave many verifiable clues. They will not cry out for help or respond to shouts—only 1% are responsive. Some dementia subjects could be viewed as "passive-evasive." Since they do not perceive themselves as lost, they would not attempt to signal or even respond to shouts. **There is a 25% fatality rate if the subject is not found within the first 24 hours.** Fatality rate is higher in hot climates and cold rainy climates. Many states are developing "Senior" or "Silver" alerts similar to AMBER alerts. Such programs are highly effective in alerting the general public.

When you find the subject, approach from the front. Make eye-contact. Non-verbal body language is highly important with dementia subjects. After assessing safety, slowly move to the subject's side. Speak slowly and in simple, concrete terms. Break down commands, questions or directions into simple, easy-to-follow components. Touching, when appropriate, is helpful. Arguing with a person with dementia is pointless and may lead to a catastrophic reaction. Instead, redirect the person with a new line of reasoning. Telling the person that a favorite person or thing is waiting for him back at base may be acceptable. Keep in mind that other impairments associated with age, such as decreased vision, hearing, and walking ability, may also be present.

Search Planning: Dementia (Alzheimer's)

Tables G-1 through G-9 are example data to support search planners in the development of search plans for persons lost and suffering from dementia.

Tables G-1 and G-2: "Distance (horizontal) from the IPP (miles and kilometers)"

Both Tables describe the crow's flight distance from the initial planning point of the search to the find location. Temperate and Dry eco-region data represent wilderness and rural cases, while Urban cases ignore the eco-region. "n" is the number of cases used in the table. The table reports the distances within which 25% of the cases were found, 50% of the cases were found, 75% and 95%. The final reported distance zone of 95% is the largest practical distance given that cases beyond 95% represent statistical outliers.

Table G-1: Distance (horizontal) from the IPP (miles)

	Temperate		Dry		Urban
	Mtn	Flat	Mtn	Flat	
n	95	175	14	15	336
25%	0.2	0.2	0.6	0.3	0.2
50%	0.5	0.6	1.2	1.0	0.7
75%	1.2	1.5	1.9	2.2	2.0
95%	5.1	7.9	3.8	7.3	7.8

Table G-2: Distance (horizontal) from the IPP (kilometers)

	Temperate		Dry		Urban
	Mtn	Flat	Mtn	Flat	
n	95	175	14	15	336
25%	0.3	0.3	1.0	0.5	0.3
50%	0.8	1.0	1.9	1.6	1.1
75%	1.9	2.4	3.1	3.6	3.2
95%	8.3	12.8	6.1	11.8	12.6

Table G-3: “Elevation (vertical) Change from the IPP (feet)”

Table G-3 breaks data out into” uphill”, “downhill” and “same elevation” in “temperate” and “dry” eco-regions. Same elevation is defined as within ± 10 feet. The top row of numbers is the percentage of subjects who traveled uphill, downhill, or stayed at the same elevation. Next are the 25%, 50%, 75%, and 95% numbers for vertical change in elevation between the Initial Planning Point and the find location. For example, in the temperate eco-region, 32% of the subjects traveled uphill and 25% of finds were within 182 feet vertical elevation change uphill relative to the Initial Planning Point.

Table G-3: Elevation (vertical) Change from the IPP (feet)						
	Temperate			Dry		
	Uphill	Down	Same	Uphill	Down	Same
%	19%	42%	39%	50%	38%	13%
25%						
50%	75	60		317	187	
75%						
95%						

Table G-4: Mobility (hours)

Table G-4 reports the amount of time the subject was moving, with data presented for “temperate” and “dry” eco-regions. “n” is the number of cases with reported data and the information is presented in 25%, 50%, 75%, and 95% quartiles. For example, in the temperate eco-region, data from 42 cases is reported with 75% of the subjects reported as moving for 3.8 hours or less. The “dry” eco-region has only 6 cases reports, so only the median or 50% quartile is reported. This data underestimates the mobility time, since finding a mobile subject will “stop” the clock.

	Table G-4: Mobility (hours)	
	Temperate	Dry
n	42	6
25%	0	
50%	0.25	4.5
75%	3.8	
100%	18	

Table G-5 “Survivability”

Table G-5 first reports the overall survivability statistics for “wilderness” and “urban” regions, then the fatality rate in 24-hour time blocks. The “<24 hours” block includes all search incidents where the subject was found within 24 hours of last being seen. The “>24 hours” block reports searches which lasted longer than 24 hours from the time the subject was last seen, the “>48 hours” block reports searches lasting longer than 48 hours from when the subject was last seen, etc. up to “>96 hours”. The survivability portion of the table answers the question: After “x” number of hours have passed, what is the probability the subject will be found alive? Looking at the table for dementia, at 48 hours after last being seen, 60% of subjects are found alive.

Table G-5: Survivability		
	Wilderness	Urban
Uninjured	73%	80%
Injured	17%	14%
Fatality	8%	6%
No Trace	2%	5%
Survivability	Alive	n
<24 hours	95%	736
>24 hours	77%	79
>48 hours	60%	30
>72 hours	60%	20
>96 hours	46%	13

Table G-6: Find Location

Table G-6 lists actual locations where subjects were found. Number of cases for eco-regions and urban areas are listed at the top, with percentage of finds associated with locations listed below. For example, the temperate eco-region includes 207 cases with 18% of finds along a road.

Table G-6: Find Location			
	Temp	Dry	Urban
n	207	7	223
Structure	20%	29%	35%
Road	18%	14%	36%
Linear	9%		9%
Drainage	9%	14%	4%
Water	7%	14%	6%
Brush	6%		1%
Scrub			
Woods	17%		3%
Field	14%	29%	6%
Rock			

Table G-7: Scenario (%)

The field lists what reportedly caused the subject to be missing. Number of incidents is listed at the top; percentages for each scenario are listed in the field. For example, 1050 incidents were considered and in 96% of the cases the subject became lost.

Table G-7: Scenario (%)	
n	1050
Avalanche	
Criminal	
Despondent	
Evading	
Investigative	2%
Lost	96%
Medical	
Drowning	
Overdue	1%
Stranded	
Trauma	

Table G-8: Track Offset (meters)

The term "track offset" as used in Table G-8 refers to the shortest perpendicular distance from the closest linear feature to where the subject was found. The track refers to any linear feature and not just the intended route of the subject. In the table which represents 110 incidents, 25% of the subjects were found within a zone 8 meters wide centered on a linear feature (a track offset of 4 meters to both sides of a linear feature).

Table G-8: Track Offset (meters)	
n	110
25%	4
50%	15
75%	71
95%	307

Table G-9: Dispersion Angle (degrees)

Table G-9 requires an incident with an Initial Planning Point and an intended destination to determine the subject's initial direction of travel or a substantial clue that provides direction of travel. Comparing the direction from the Initial Planning Point to the find location with the original direction of travel yields a dispersion angle in degrees.

Table G-9 reports on 11 cases with data available (n), and shows the degrees of dispersion for 25%, 50%, 75%, and 95% quartiles.

For example, in 50% of the incidents, the subject was found within 23 degrees of the subject's initial direction of travel.

	Table G-9: Dispersion Angle (degrees)
	Temperate
n	11
25%	11
50%	23
75%	66
100%	70

Dementia (Alzheimer's) Initial Reflex Tasking Actions

Figure G-1 on the next page provides suggested Initial Reflex Tasking Actions for lost persons suffering from dementia (Alzheimer's).

1. Investigation

- [ ] Determine planning data.
- [ ] Determine searching data.
- [ ] Start MPQ/LPQ. Ask specific questions.
- [ ] Previous wandering.
- [ ] Potential destinations.
- [ ] Severity of dementia.
- [ ] Check taxis, mass transit (all shifts), hospitals, EMS, jails, shelters, etc.
- [ ] Alert municipal workers.
- [ ] Issue "Silver" alert.

2. Initial Planning Point

- [ ] Preserve IPP.
- [ ] Locale search.
- [ ] Highly systematic grounds and structure search.
- [ ] Repeat search of grounds and structure.
- [ ] Task sign cutters/trackers.
- [ ] Tracking/trailing dogs; ideal if they have practiced with persons with dementia.

3. Hub/Immediate Area

- [ ] Canvass campground.
- [ ] Canvass neighborhood.
- [ ] Thorough search of 25% zone. Sweep/area dogs.
- [ ] Notify community by media, flyer, door to door, and/or telephone system.

4. Containment

- [ ] Establish containment.
- [ ] Use statistical max zone or theoretical zone.
- [ ] Containment provided by road patrols/air.
- [ ] In urban environment, use road, bike, air patrols.
- [ ] Contain entrances to gated developments.

5. Travel Corridors

- [ ] Hasty search of trails, roads, drainages, and other routes leading away from IPP, patrol roads.
- [ ] Look for decision points and cut for sign at turn off points.
- [ ] Dogs into drainages.
- [ ] Corridor search parallel to roads and routes.

6. High Probability Tasks

- [ ] Check historical finds.
- [ ] Previous lost locations.
- [ ] High hazard areas.
- [ ] Previous homes.
- [ ] Ensure heavy brush is searched.

Figure G-1: Dementia (Alzheimer's) Initial Reflex Tasking Actions

Dementia (Alzheimer's) Additional Investigative Questions

- What is the exact diagnosis of the type of dementia?
- Name and contact number of neurologist/gerontologist.
- Last Mini-Mental Status Exam (MMSE), if known.
- Describe subject as mild, moderate, or severe dementia.
- Determine which description is most appropriate:
- Mild confusion and forgetfulness, short-term memory affected.
- Difficulty distinguishing time, place and person. Some language difficulties.
- Nearly complete loss of judgment, reasoning, and loss of some physical control.
- Does the subject know his own name?
- Does the subject know where he is when at home?
- Does the subject recognize the local neighborhood?
- Does the subject recognize familiar faces?
- Will the subject answer to his name being called?
- Is the subject able to conduct a conversation?
- How long can the subject do or discuss something before forgetting?
- How long does a conversation last until an average person suspects something
- is wrong or not quite right?
- Describe the subject's ability to tell time.
- Has the subject experienced personality or emotional changes? Describe.
- Does the subject have delusions? Describe.
- Does the subject have paranoia? Describe.
- Does the subject have hallucinations? Describe.
- Does the subject have depression? Describe.
- Has the subject experienced an emotional breakdown? Describe.
- Has the subject shown violence towards others? Describe.
- Is the subject registered in the Alzheimer's Association's MedicAlert® + Safe Return® program or any other similar registry? Describe any identification or marking jewelry or labels the subject might be wearing.
- List all of the subject's addresses, dwelling types, and how long he has lived at each address going back to childhood. List locations even if they no longer exist. What jobs and occupations did the subject have at each location?
- Did the subject recently move or change locations? If so, when? What was the previous location? How has he adjusted to being in a nursing home/new location?
- How have caregivers adjusted their routines?
- List of all immediate relatives and distant relatives the subject communicated with during their lifetime?
- Is the subject familiar with the area where last seen?

- What is the subject's favorite place?
- Has the subject been involved with outdoor classes, scouting, military, overnight experiences, or outdoor recreation? Describe.
- Is the subject afraid of noises, crowds, dogs, traffic, water, horses, the dark, or other items? Describe how he reacts.
- How does the subject respond to strangers? Does he approach strangers?
- Is the subject dangerous to himself? Dangerous to others?
- Has the subject ever wandered away before, and you (the interviewee) did not know his location for 5 minutes or more? If so, for each incident describe the following: Where was the subject last seen? What was the subject doing when last seen? Events that might have caused the subject to have wandered? What actions were taken? Where was the subject found? What was the distance "as the crow flies" from the point the subject was last seen and where found? How was the subject found? List any medical problems that resulted from being lost.
- What jobs and occupations has the subject held throughout his life?
- Subject's hobbies and interests? Able to still engage in hobby?
- What are the subject's daily habits? Did they occur on the day last seen?
- Distance subject typically walks each day (e.g., during the past week).
- Number of walks during the past week.
- Greatest distance subject has walked during the past three months? During the past ten years?
- Estimate the greatest distance you believe the subject could walk.
- Describe the subject's ability to walk including any unique gait or shuffle. List any limitations to walking.
- Does the subject talk about a person or place? Is it out of town? Describe.
- Does the subject talk about a person who is no longer alive?
- Does the subject talk about visiting a person or place that is out of town? Describe.
- Has the subject attempted to visit a person or place without supervision? Describe.
- Can the subject drive a car safely? Can the subject find keys and start a car? Does he still have a driver's license? When was the last time he drove or expressed a desire to drive?
- Does the subject travel independently using public or private transportation?
- Has the subject attempted to travel independently on public or private transportation in the last six months?
- Does the subject walk or travel a considerable distance from home and return unaided? Describe.
- Does the subject get lost or confused easily in an unfamiliar setting?
- Does the subject get lost or confused easily at home or in living quarters?
- Does the subject wander? Wander at night? Wander during the day?
- Does the wandering appear goal-directed? Describe.
- Does the wandering appear random?
- Does the subject seek out exits or try to escape from present location?

- Does the wandering appear related to a search for a person or place? Provide a timeline of all events during the day.
- What was his emotional state when he was last seen? How does that compare to his baseline?
- Determine the door the subject most likely exited from and attempt to determine a direction of travel.
- How old is the provided picture? Describe any changes since the photograph was taken? How old does the subject actually appear now? Are any recent videotape or DVD images available?

Appendix H: Requests for Federal SAR Assistance

Search for Missing Persons

Posse Comitatus Act Support Limitations

Rescue

Patient Transfer

Mercy Missions

SAR Facility Transport

Federal RCCs require the following information to properly evaluate a request for assistance by the responsible agency concerning missing person searches, rescues, patient transport, mercy missions, and SAR facility transports.

Search for Missing Persons

- Mission Objective;
- Objective Description: Name, Age, Sex, Clothing, City and State of Residence;
- Known Medical Problems, including physical and mental conditions at the time of event;
- Known survival gear/experience;
- Type of activity engaged in when event occurred;
- Date time group and location of last known position;
- Destination and time expected to arrive;
- Current Search Area;
- Type of assistance requested (air and/or ground resources);
- Other SAR forces involved;
- Current weather; and
- Incident commander/On Scene Coordinator contact information.

Posse Comitatus Act Support Limitations

Requests for DoD SAR assistance cannot violate the *Posse Comitatus Act*. The "YES" answer to the following questions may constitute a Posse Comitatus situation and should be reviewed with legal counsel.

- Is the objective the subject of a criminal investigation?
- Is the objective suspected of being involved in criminal activity?
- Is the objective a fugitive from the law and/or want to be found?
- Does the objective pose a danger to the search teams?

Rescue

- Mission Objective;
- Objective Description: Name, Age, Sex, Clothing, City and State of Residence;
- Nature of emergency (injuries, known medical issues, location, terrain, etc.);
- Known survival gear/experience;
- Rescue location (lat/long);

- Destination location (lat/long);
- Type of assistance requested (air and/or ground resources);
- Other SAR forces involved;
- Current weather; and
- Incident commander/On Scene Coordinator contact information

Patient Transfer

A patient transfer is the movement of patient(s) from one medical facility to another.

- Mission objective;
- Objective's description: name, age, sex, city, and state of residence;
- Nature of emergency;
- Point of origin;
- Destination; and
- Type of assistance requested (air or ground resource).

Mercy Missions

A mercy mission is the transport of life-saving materials (e.g., blood, organs, the donor, etc.) to save life.

- Objective description: object being transported and name, age, sex of the person for whom the object is being transported (e.g. the organ recipient);
- Nature of emergency (reason for transport);
- Pickup location;
- Destination;
- Type of assistance requested (air or ground); and
- Point of contact information for material to be transported.

SAR Facility Transport

- Mission objective
- Objective's description: name, age, sex, city, and state of residence
- Nature of emergency
- Point of origin
- Destination
- Type of assistance requested (air or ground resource)
- Incident commander/On Scene Coordinator contact information

RCCs cannot guarantee return transport, but will make every effort to provide transport if the requesting agency cannot. Just as the requesting agency must exhaust all local and state resources before asking any RCC for assistance on scene, the requesting agency must make the same effort for the return trip.

Appendix I: National SAR Plan of the United States

National Search and Rescue Plan of the United States

POLICY

1. It is the policy of the signatory federal agencies to provide a *National Search and Rescue Plan* (referred to as the "Plan") for the United States for coordinating search and rescue (SAR) services to meet domestic needs and international commitments. Implementing guidance for this Plan is provided in the *International Aeronautical and Maritime Search and Rescue Manual* (IAMSAR Manual discussed below), the *National Search and Rescue Supplement* (NSS)[1], and other relevant directives of the Participants to this Plan.

PURPOSE

2 This Plan continues, by interagency agreement, the effective use of all available resources in all types of civil SAR missions to enable the United States to satisfy its humanitarian, and national and international legal obligations.

3 The *National Search and Rescue Plan*-1999 is superseded by this Plan.

TERMS AND DEFINITIONS

4. The following terms and definitions are used in this Plan.[2]

a. Civil Search and Rescue (civil SAR): Search operations, rescue operations, and associated civilian services provided to assist persons and property in potential or actual distress in a non-hostile environment;

b. Incident of National Significance: An actual or potential high-impact event that requires robust coordination of the Federal response in order to save lives and minimize damage, and provide the basis for long-term community and economic recovery;

c. International Civil Aviation Organization (ICAO): A specialized agency of the United Nations that promotes the safe and orderly development of international civil aviation, and provides for establishment of aeronautical SAR services;

d. Incident Command System (ICS): A standardized on scene emergency construct specifically designed to provide for the adoption of an integrated organizational structure that reflects the complexity and demands of single or multiple incidents, without being hindered by jurisdictional boundaries;

---

[1] The NSS provides guidance to federal agencies concerning implementation of this Plan, builds upon the baseline established in the IAMSAR Manual and encourages each SAR agency to develop an addendum to the NSS that provides agency specific direction and guidance for implementing the NSS and IAMSAR Manuals.

[2] Additional terms and definitions are available in the IAMSAR Manual and the National Response Plan. (Note: The National Response Plan has been replaced by the National Response Framework (NRF).

e. International Maritime Organization (IMO): A specialized agency of the United Nations that promotes the safety and security of international shipping, and provides for establishment of maritime SAR services;

f. Mass Rescue Operations (MROs): SAR services characterized by the need for immediate response to large numbers of persons in distress, such that the capabilities normally available to SAR authorities are inadequate;

g. National Incident Management System (NIMS): An ICS that provides a consistent, nationwide approach for federal, state, local, and tribal governments; the private sector; and nongovernmental organizations to work effectively and efficiently together to prepare for, respond to, and recover from domestic incidents, regardless of cause, size or complexity;

h. National Response Plan (NRP): The federal government's guiding document that establishes a comprehensive all-hazards approach to enhance the ability of the United States to manage domestic incidents;

i. National Search and Rescue Committee (NSARC): The interagency Committee that oversees the NSP and serves as a federal coordinating forum for national civil SAR matters;

j. Participant: A signatory federal agency to this Plan;

k. Rescue: An operation to retrieve persons in distress, provide for their initial medical or other needs and deliver them to a place of safety;

l. Rescue Coordination Center (RCC): A unit, recognized by ICAO, IMO or other cognizant international body, responsible for promoting efficient organization of SAR services and for coordinating the conduct of SAR operations within a Search and Rescue Region (SRR);

m. Rescue Sub-center (RSC): A unit subordinate to an RCC established to complement the latter according to particular provisions of the responsible authorities;

n. Search: An operation using available personnel and facilities to locate persons in distress;

o. Search and Rescue Coordinator: A federal person or agency with overall responsibility for establishing and providing SAR services for a search and rescue region(s) for which the U.S. has primary responsibility;

p. Search and Rescue Region (SRR): An area of defined dimensions, recognized by IMO or ICAO, associated with a rescue coordination center, within which SAR services are provided;

q. Search and Rescue Services: The performance of distress monitoring, communication, coordination and SAR functions, including provision of medical advice, initial medical assistance, or medical evacuation, through the use of public and private resources including cooperating aircraft, vessels and other craft and installations;

r. Search and Rescue Satellite-aided Tracking (SARSAT): A national and international satellite Program that helps to protect life and property by providing accurate, timely, and reliable distress and security alert and location information to SAR authorities worldwide and that, along with the comparable Russian Cospas Program, comprises the international Cospas-Sarsat satellite system; and

s. Urban Search and Rescue (US&R): The location, rescue (extrication), and initial medical stabilization of victims trapped in confined spaces.

OBJECTIVES

5. Knowing the importance of cooperation in providing expeditious and effective SAR services, the Participants desire to:

   a. Provide a United States Plan for coordinating civil SAR services to meet domestic needs, international commitments, and to document related national policies;

   b. Support lifesaving provisions of IMO's International Convention on Maritime Search and Rescue, ICAO's Convention on International Civil Aviation (Annex 12), certain international agreements to which the United States is Party, and similar international instruments;

   c. Provide an overall Plan for coordination of civil SAR operations, effective use of available resources, mutual assistance, and efforts to improve such cooperation and services;

   d. Integrate available civil SAR resources into a cooperative network for greater protection of life and property and to ensure greater efficiency and economy; and

   e. Enable the United States to satisfy its humanitarian, and national and international legal obligations.

6. This Plan is further intended to:

   a. Provide national guidance for development of civil SAR-related systems;

   b. Describe its Participants and their roles in a pro-lifesaving context;

   c. Recognize lead federal agencies, respectively, for the types of operations covered by this Plan, and describe civil SAR responsibilities;

   d. Account for saving property, but on a secondary basis to saving lives;

   e. Account for all operations up to and including providing initial assistance (food, clothing, medical, etc.) to SAR survivors and delivering them to a place of safety; and

   f. Have, as a primary concept, cooperation for overall and continual development, coordination and improvement of civil SAR services.

SCOPE

7 It is intended that this Plan not conflict in any way with civil SAR responsibilities agreed to by contracting States of the Convention on International Civil Aviation, the International Convention on Maritime Search and Rescue, or other appropriate international instruments to which the U.S. is or may become a Party.

8 No provisions of this Plan or any supporting plan are to be construed in such a way as to contravene responsibilities and authorities of any Participant as defined by statutes, executive orders or international agreements, or of established responsibilities of other agencies and organizations which regularly assist persons and property in distress resulting from incidents of a local nature

9 This Plan is solely intended to provide guidance to the Participants. State authorities may retain SAR responsibilities within their boundaries for incidents primarily local or intrastate in character. In such cases, appropriate agreements are generally made between federal civil SAR coordinator(s) and relevant State authorities.

NATIONAL SAR COMMITTEE

10. The National Search and Rescue Committee (NSARC) is responsible for the provisions of this Plan, consistent with applicable laws and executive orders, and coordinates and provides guidance for its implementation.

PARTICIPANTS

11. The Participants to this Plan correspond to the NSARC Member Agencies, and are as follows:

a. Department of Homeland Security (DHS) carries out responsibilities to protect against and respond to hazards and distress situations affecting the nation and its people;

(1) United States Coast Guard develops, establishes, maintains and operates civil SAR resources for the promotion of safety on, under and over international waters and waters subject to United States jurisdiction. The Coast Guard has specialized expertise, facilities, and equipment for responding to maritime distress and other situations; and

(2) Federal Emergency Management Agency (FEMA) coordinates federal response under the guidelines of the NRP and oversees the national US&R response system.

b. Department of Transportation (DOT) carries out broad responsibilities in transportation safety;

(1) Federal Aviation Administration (FAA) establishes and enforces flight safety regulations and operates the air traffic control, navigation, and flight service facilities that are available to assist in civil SAR operations; and

(2) The Maritime Administration (MARAD) maintains a fleet of ready reserve vessels for government use in contingencies and supports and promotes commercial maritime safety with regard to vessels, equipment, and mariners. MARAD advocates efficient maritime intermodal commerce through enhanced security.

c. Department of Defense (DOD) components have facilities and other resources that are used to support their own operations. These resources may be used for civil SAR needs to the fullest extent practicable on a non-interference basis with primary military duties according to applicable national directives, plans, guidelines and agreements. Within DOD, the United States Air Force provides and uses resources for the efficient organization and coordination of civil SAR services, including SAR operations, within its assigned SRR;

d. Department of Commerce (DOC) participates in or supports civil SAR operations through the National Oceanic and Atmospheric Administration (NOAA). NOAA provides satellite services for detecting and locating persons in potential or actual distress. NOAA is the lead federal agency for SARSAT and associated international Programs, including Cospas-Sarsat. NOAA also provides nautical and aeronautical charting; information on tides and tidal currents; marine environmental forecasts and warnings for the high seas as well as coastal waters and inland waterways, as well as environmental data used for search planning for civil SAR operations;

e. Federal Communications Commission (FCC) promulgates rules and regulations for non-government use of wire and radio facilities for promoting safety of life and property, and cooperates in civil SAR operations through its long-range direction finder network;

f. The National Aeronautics and Space Administration (NASA) supports civil SAR objectives through research and development or application of technology to search, rescue, survival, and recovery systems and equipment, such as location tracking systems, transmitters, receivers, and antennas capable of locating aircraft, ships, spacecraft, or individuals in potential or actual distress. Additionally NASA has aircraft, spacecraft and worldwide tracking, data acquisition and communications networks that can assist in civil SAR operations; and

g. The Department of the Interior (DOI) National Park Service (NPS) provides civil SAR services on lands and waters administered by NPS, may assist federal, State, local and tribal authorities with emergency response operations outside NPS jurisdictions, and supports response operations during Incidents of National Significance.

12. A federal agency that is not a Participant may become a Participant in accordance with the procedures set forth in the *United States National Search and Rescue Committee Interagency Agreement.*

UNITED STATES SAR REGIONS

13. SRRs are established to ensure provision of adequate land-based communications infrastructure, efficient distress alert routing, and proper operational coordination to effectively support civil SAR services.
14. SRRs should be contiguous and, as far as practicable, not overlap.
15. Establishment of SRRs is intended to effect an understanding concerning where nations have accepted primary responsibility for coordinating or providing civil SAR services. The existence of SRR limits should not be viewed as a basis to restrict, delay, or limit in any way, prompt and effective action to relieve distress situations.
16. All SRRs of the United States are established in cooperation with neighboring nations, are internationally recognized, and are described in pertinent IMO or ICAO documents.[3]
17. United States maritime and aeronautical SRRs will be harmonized with each other to the extent practicable, recognizing, however, that lines separating SRRs must normally be agreed by governments having neighboring SRRs when possible. SRRs will not be allowed to unduly affect or be affected by any political boundaries.
18. As defined in the Convention on International Civil Aviation (Annex 12) and the International Convention on Maritime Search and Rescue, one RCC is associated with each recognized SRR. Comprehensive IMO and ICAO standards and guidance pertinent to these RCCs have been developed and may be found in relevant Conventions, the IAMSAR Manual, and other publications which should be held and used by United States RCCs.

3 United States maritime and aeronautical SRRs are established in accordance with the relevant IMO and ICAO Conventions and with the guidance of the IAMSAR Manual. These SRRs are internationally-recognized and documented in the appropriate ICAO Regional Air Navigation Plans and in the IMO SAR Plan. More specific information on United States SRRs can also be found in the NSS in which SRR charts are included for convenient reference.

19. United States SAR Coordinators, as designated in this Plan, are responsible for arranging civil SAR services and establishing RCCs for their respective SRRs; these RCCs must be staffed with trained personnel on a 24-hour basis. The United States civil SAR system becomes integrated into the global SAR system by establishing recognized SRRs and RCCs which comply with international standards. The geographic areas for which civil SAR Coordinators are responsible are assigned in the Section of this Plan on "SAR Responsibilities" below. SAR Coordinators are responsible for all the types of civil SAR services covered by this Plan (see paragraph 45) relevant to their respective SRRs.[4]
20. An SRR may be subdivided as long as the delimitation of the sub-regions coincides with the SRR limits. Where this is not practicable, changes to the SRR limits should be coordinated with any affected neighboring nation, and for aeronautical or maritime SRRs, the changes must be recognized by ICAO or IMO, respectively, in accordance with requirements of those Organizations.

SAR RESONSIBILITIES

United States SAR Coordinators

21. The SAR Coordinators, designated below, have overall responsibility for establishing RCCs as necessary, and for providing or arranging for SAR services within United States SRRs. Only RCCs properly established by these SAR Coordinators should carry out domestic and international coordination of civil SAR operations.
22. United States SAR Coordinators are listed below:

   a. United States Air Force: Recognized SAR Coordinator for the United States aeronautical SRR corresponding to the continental United States other than Alaska;
   b. United States Pacific Command: Recognized SAR Coordinator for the United States aeronautical SRR corresponding to Alaska; and
   c. United States Coast Guard: Recognized SAR Coordinator for all other United States aeronautical and maritime SRRs. This includes the State of Hawaii as well as waters over which the United States has jurisdiction, such as navigable waters of the United States.

Other Responsibilities

23. National Park Service. The NPS provides emergency services on lands and waters administered by NPS, assists visitors within the National Parks or National Monuments, and aids authorities in neighboring jurisdictions. Civil SAR operations, including emergency medical aid, are conducted in a wide variety of environments such as remote, rural and roadless areas, lakes, rivers and oceans, and deserts, mountains and caves, and often require extended response times and use of specialized equipment. The NPS works closely with State, local and volunteer SAR organizations, and is proficient in use of the incident command system.

[4]Aeronautical SRRs are documented in appropriate ICAO Regional Air Navigation Plans, and maritime SRRs are documented in the IMO Global SAR Plan; the SRRs of the United States are also depicted in the NSS

24. Critical Support Functions. SAR services include distress alerting and other critical functions that could not be carried out without the support of NASA, NOAA, the FCC and other entities. While most of this Plan concerns matters directly related to SAR operations, the importance of support functions of such Agencies cannot be over-stated. Close cooperation and mutual support among all the Participants to this Plan are vital to conducting effective SAR operations.
25. Federal Emergency Management Agency: FEMA oversees the national US&R response system.
26. Other Assistance: The Participants to this Plan are encouraged to support each other and other government authorities for civil SAR or other emergency response operations in accordance with applicable legislative authority.
27. Incidents of National Significance: Participants may use their specialized capabilities to support civil SAR or other operations associated with an Incident of National Significance when the NRP is implemented with lifesaving efforts normally having priority over all other activities.
28. State and local responsibilities: Outside the above listed national SAR Coordinator and federal civil SAR responsibilities, State and local authorities are responsible for land based SAR and designate a person to be "SAR Coordinator" within their respective jurisdictions. State SAR Coordinators are integral partners of the national SAR Coordinators and are critical to providing effective civil SAR services.

International Representation

29. The Department of State normally designates federal agencies responsible for civil SAR to represent the United States in the following international forums:
    a. The Coast Guard leads and coordinates national participation in IMO's SAR and safety-related initiatives;
    b. The Federal Aviation Administration leads and coordinates national participation in ICAO's safety-related initiatives;
    c. NOAA leads and coordinates national participation in SARSAT and associated international programs; and
    d. Based upon invitations from ICAO and IMO, respectively, the Air Force will provide an aeronautical SAR expert and the Coast Guard will provide a maritime SAR expert to serve as members of the ICAO-IMO Joint SAR Working Group.

Support Outside United States SAR Regions

30. SAR Coordinators, as well as other United States authorities, may support civil SAR operations anywhere in the world, consistent with their expertise, capabilities and legal authority. This is consistent with the principles of assisting persons in distress without regard to nationality or circumstances and of using all available resources for SAR. It is in the interest of the safety of United States citizens who travel or live worldwide. It is also consistent with United States humanitarian goals and the advantages of domestic and international cooperation.
31. In accordance with international law, United States SAR facilities, in a position to render timely

and effective assistance, may enter into or over the territorial seas or archipelagic waters of another state for the purposes of rendering assistance to a person, ship, or aircraft whose position is reasonably well known, is in danger or distress due to perils of the seas, and requires emergency assistance.

32. Participants, consistent with their capabilities and legal authority, will support another country's SAR operations in territorial and international waters beyond recognized United States aeronautical and maritime SRRs. As appropriate, and within their capabilities, DOD combatant commanders should provide such support within their respective geographic areas of responsibility.

33. In carrying out civil SAR support functions with other nations, such as training, exercises, and liaison, each Participant will coordinate its activities with other Participants having SAR expertise with respect to the support concerned.[5]

34. To the extent the Participants have the authority to do so, they may maintain liaison and cooperate with authorities of other nations that have comparable responsibilities for providing civil SAR services. Such cooperation should be carried out in coordination with the United States SAR Coordinators, and with other neighboring SAR authorities that might be affected, as appropriate. When such coordination deals substantially with matters relating to actual conduct of civil SAR operations, it should normally include United States Coast Guard Headquarters, Office of Search and Rescue, in order to ensure consistency with United States obligations under international agreements to which the United States is a Party, and compliance with the IAMSAR Manual and other international guidance relevant to implementing such agreements.

35. Participants should not accept a SAR Coordinator or RCC role for civil SAR operations for SRRs for which other nations are responsible. However, the Participants may provide and support civil SAR operations in such areas when:

   a. Assistance is requested (normally this should be in accordance with RCC-to-RCC procedures prescribed in the IAMSAR Manual);
   b. U.S. citizens are involved; or
   c. U.S. facilities become aware of a distress situation to which no other suitable facilities are responding, or where other available SAR services appear to be inadequate.

36. For distress situations in international waters or airspace where no SRR exists for which an RCC is responsible, or where it appears that the responsible RCC is not responding in a suitable manner, United States RCCs or facilities will assist as appropriate. Such assistance will be subject to availability of resources, legal constraints, and other applicable United States policies.

---

[5] A wealth of valuable reference material is available which should be used working with other nations in the area of SAR. These include, but are not limited to, the SAR-related conventions, the IAMSAR Manual (three volumes), this Plan, the NSS, information about the AMVER ship reporting system, and many documents of Cospas-Sarsat, IMO, ICAO, etc. Some of these references are available in languages other than English. Participants should be familiar with such references, and use them as appropriate.

37. Provisions of international conventions dealing with SAR are intended to ensure that suitable civil SAR services are in place and responsibilities are assigned worldwide to assist persons in danger or distress. However, there may be nations which are not Parties to, or which have not yet fully complied with, these conventions. Therefore, distress situations may arise for United States resources to supplement civil SAR capabilities in certain geographic areas. Additionally, the United States assistance may support these nations and help develop their SAR capabilities by providing training or other means, consistent with United States law. Participants may take advantage of such situations as appropriate.

38. When assisting other nation's SAR authorities, or other agencies and organizations that support these SAR authorities, Participants should ensure:
   a. They have appropriate legal authority and expertise to do so;
   b. Principles or provisions of conventions or agreements to which the United States is Party are not violated;
   c. Applicable procedures set forth in the IAMSAR Manual, NSS, and other relevant directives are known and followed;
   d. Such efforts are carried out in consultation with other Participants as appropriate; and
   e. The authorities assisted are responsible for the civil SAR functions in that country.

39. Policies on rendering assistance in foreign territories or territorial waters must balance concerns for saving lives, sovereignty and national security. Provisions for territorial entry, as necessary, should be addressed in international civil SAR agreements; care should be taken to ensure that such agreements are compatible with national policies in this regard.

40. When any Participant to this Plan is addressing civil SAR-related inquiries or proposals from other nations or organizations outside the United States, or when hosting or attending international SAR meetings, care should be taken that interested United States agencies, organizations, or persons are consulted and involved as appropriate.

SAR AGREEMENTS

41. Bilateral or multilateral SAR agreements with other United States agencies or organizations, or with international authorities or organizations of other nations, may be of practical value for SAR by:
   a. Helping to fulfill United States domestic or international obligations and needs;
   b. Enabling more effective use of all available SAR resources;
   c. Better integration of United States civil SAR services within the global SAR system;
   d. Building mutual commitment to support civil SAR;
   e. Resolving civil SAR procedures and sensitive matters in advance of time-critical distress situations; and
   f. Identifying types of cooperative matters and efforts which may enhance or support civil SAR operations (i.e., access to medical or fueling facilities, training and exercises, meetings, information exchanges, use of communications capabilities and joint research and development projects, etc.).

42. Negotiation and conclusion of such agreements should consider matters such as:
   a. Which authorities of the Government, agencies, or organizations concerned should be involved with the agreement;
   b. Which types of civil SAR operations (e.g., aeronautical, maritime, etc.) or SAR support functions should be included within the scope of the agreement;
   c. Consistency with international and domestic civil SAR principles or policies;
   d. Establishment of lines separating SRRs if relevant;
   e. Whether other treaties, agreements, etc., exist which should be superseded or accounted for in preparation of a new agreement; and
   f. Relevant guidance of the IAMSAR Manual, NSS, and other pertinent directives.

43. Participants which develop any agreement concerning civil SAR shall ensure that such efforts are coordinated with other interested Participants.

44. Any such international agreement may not be signed or otherwise concluded without prior consultation with the Secretary of State (refer to Title 1 USC 112b).

SAR SERVICES COVERED BY THIS PLAN

45. This Plan covers the following types of civil SAR services:
   a. Maritime (involving rescue from a water environment);
   b. Aeronautical (including civil SAR assistance in the vicinity of airports);
   c. Land (including civil SAR operations associated with environments such as remote areas, swift water, caves, mountains, etc.);
   d. Urban search and rescue (US&R);
   e. Provision of initial assistance at or near the scene of a distress situation (e.g., initial medical assistance or advice, medical evacuations, provision of needed food or clothing to survivors, etc.);
   f. Delivery of survivors to a place of safety or where further assistance can be provided, or further transportation arranged if necessary;
   g. Saving of property when it can be done in conjunction with or for the saving of lives;
   h. Mass rescue operations (MROs); and
   i. SAR services associated with Incidents of National Significance covered by the National Response Plan (NRP).[6]

---

[6] The NRP covers Federal responses to declared Incidents of National Significance; this Plan covers all civil SAR operations, whether conducted independently or concurrently with the NRP. If carried out concurrently, this Plan becomes an NRP supporting plan by integrating on matters relating to coordination and conduct of disaster response SAR operations. When the NRP is implemented, civil SAR operations will continue to be covered by this Plan; the only difference is that civil SAR services will be coordinated with other NRP operations on scene via the Operations Section of the Incident Command Post or Unified Command established under NIMS. During NRP implementation, SAR facilities may be tasked to support other operations carried out under the NRP on a not-to interfere basis with lifesaving efforts, or assistance for SAR services may be supplemented by resources available under the NRP. *(Note: The NRP was replaced by the NRF in 2008)*

46. This Plan does ***not*** cover operations such as:

a. Air ambulance services which did not result from a rescue or recovery operation;

b. Rescues from space (although rescue of persons returned from space can be included);

c. Military operations, such as combat SAR or other types of recovery by military operations to remove military or civilian personnel from harm's way;

d. Salvage operations;

e. Assistance in cases of civil disturbance, insurrection or other emergencies which endanger life or property or disrupt the usual process of government; and

f. Operations and coordination in addition to those covered by this Plan that might be carried out concurrently with civil SAR operations on scene, such as could occur during a disaster or terrorism response situation, or an Incident of National Significance.

MUTUAL ASSISTANCE

47. The Participants agree to cooperate as follows:

a. Provide mutual agency support by pooling relevant facilities and support services as appropriate for operations within their respective SRRs, and consistent with each participant's relevant legal authorities;

b. Make and respond to requests for operational assistance between RCCs, RSCs, or comparable Participant command centers (CCs) as capabilities allow;

c. Develop procedures, communications, and databases appropriate for coordination of facilities responding to distress incidents and for coordination between the RCCs, RSCs or Participant CCs;

d. Normally follow applicable IMO, ICAO, or other relevant international guidance regarding operational procedures and communications; and

e. In areas where more than one authority may respond to distress situations, agreed procedures should be in place, which balance concerns for saving lives and for jurisdiction.

48. Participants may enter into other collaborative efforts with each other such as:

a. Mutual visits, information exchanges, and cooperative projects for support of civil SAR;

b. Joint training or exercises;

c. Cooperation in development of procedures, techniques, equipment, or facilities;

d. Establishment of groups subordinate to NSARC as a means for more in-depth focus on matters of common concern; and

e. Carry out cooperative efforts similar to those indicated above on an international level.

GENERAL TERMS

49. No provision of this Plan or any supporting plan is to be construed as an obstruction to prompt and effective action by any agency or individual to assist persons in distress.

50. Cooperative arrangements between Participants with operational responsibilities and state, local,

and private agencies should provide for the fullest practicable mutual cooperation for civil SAR operations, consistent with the willingness and ability of such agencies to act, and for such coordination by the responsible RCC, RSC, or CC of their facilities as may be necessary and practicable.

51. Participants with operational responsibilities may request assistance from other federal agencies having capabilities useful for a civil SAR mission.
52. The federal government does not compel state, local or private agencies to conform to this Plan; such entities can direct and control their own facilities within their boundaries, and cooperation will be pursued through liaison and consultation.

CHARGING FOR SAR SERVICES

53. Each Participant will fund its own activities in relation to this Plan unless otherwise provided for by law or arranged by the Participants in advance, and will not allow cost reimbursement to delay response to any person in danger or distress.[7]
54. Participants agree that unless required for by law, civil SAR services provided to persons in danger or distress will be without subsequent cost-recovery from the person(s) assisted.
55. In accordance with customary international law, when a nation requests help from another nation to assist person(s) in danger or distress, if such help is provided, it will be accomplished voluntarily; the United States will neither request nor pay reimbursement costs for such assistance.

GUIDING PRINCIPLES

General

56. Participants coordinating operations should, consistent with applicable laws and executive orders, organize existing agencies and their facilities through suitable agreements into a basic network to assist military and non-military persons in actual or potential distress, protect property and to carry out obligations under customary international law and international instruments to which the United States is a Party.
57. Participants will seek to keep political, economic, jurisdictional, or other such factors secondary when dealing with lifesaving matters.
58. Consistency and harmonization will be fostered wherever practicable through plans, procedures, equipment, agreements, training, terminology, etc., for the various types of lifesaving and recovery operations, taking into account terms and definitions adopted internationally as much as possible.
59. Terminology and definitions used throughout the United States civil SAR community will be standardized to the extent possible, and be as consistent as possible with usage in pertinent international conventions and the IAMSAR Manual.

---

[7] In some cases, SAR operations may be reimbursed under provisions of the Stafford Act when the NRP is implemented, and may include civil SAR activities leading up to implementation of the NRP; however, such funding provisions are beyond the scope of this Plan, and possibility of such reimbursement should never adversely affect the commencement or conduct of needed civil SAR operations. (*Note: NRP replaced by NRF*)

60. If a distress situation appears to exist or may exist, search, rescue or similar recovery efforts will be based on the assumption that a distress situation does actually exist until it is known differently.
61. Assistance will be provided to persons in distress without regard to their nationality, status, or circumstances.
62. Generally, cost-effective safety, regulatory, or diplomatic measures that tend to minimize the need for United States services will be supported.

63. Close cooperation will be established between services and organizations, which may support improvements in lifesaving functions in areas such as operations, planning, training, exercises, communications and research and development.

64. Recognizing the critical importance of reduced response time in successful rescue and similar efforts, a continual focus will be maintained on developing and implementing means to reduce the time required for:

    a. Receiving alerts and information associated with distress situations;
    b. Planning and coordinating operations;
    c. Facility transits and searches;
    d. Rescues; and
    e. Providing immediate assistance, such as medical assistance, as appropriate.

65. All aeronautical and maritime civil SAR personnel should be generally familiar with IMO's International Convention on Maritime Search and Rescue, ICAO's Convention on International Civil Aviation, Annex 12 ("Search and Rescue"), the joint ICAO-IMO IAMSAR Manual, the NSS, and other primary directives or applicable information.
66. Local cooperative arrangements within the United States should be made in advance between civil SAR, air traffic, and airport authorities for close coordination in handling aircraft emergencies, unless the same authorities hold all the involved responsibilities.
67. SAR principles and procedures of relevant customary international law, international Conventions and the IAMSAR Manual will serve as the framework for coordination of civil SAR operations, and especially civil SAR operations involving multiple countries, organizations or jurisdictions; U.S. organizational or operational civil SAR plans and provisions of the NSS will be consistent with these international provisions to the extent practicable.
68. Operational responsibilities for maritime and aeronautical civil SAR will generally be associated with internationally-recognized geographic maritime and aeronautical SRRs. A single federal agency will be given primary responsibility for coordinating civil SAR operations within each SRR, with other agencies and organizations providing support as appropriate. However, in some specific sub-areas, such as within national parks, other federal authorities may be responsible.
69. Distress situations involving airborne aircraft will normally be handled by the maritime or aeronautical civil SAR authorities responsible for the SRR concerned once the distressed aircraft is down; cooperation between these civil SAR authorities and air traffic service authorities will continue while the aircraft remains airborne.
70. Land civil SAR services may include aeronautical civil SAR operations. Participant's involvement in such operations may be governed by agreements between SAR coordinators and

various state and local authorities. Participants will support such operations, as appropriate, bearing in mind the provisions of paragraph 9 of this Plan.

Coordination of Operations

71. Each agency responsible for operations under this Plan will:

    a. Keep information readily available on the status and availability of key civil SAR facilities or other resources which may be needed for operations; and

    b. Keep each other fully and promptly informed of operations of mutual interest, or which may involve use of facilities of another Participant;

72. SAR Coordinators will delegate to their RCCs the authority to:

    a. Request assistance via other RCCs/RSCs including those of other nations;

    b. Promptly respond to requests for assistance from other RCCs/RSCs, including those of other nations as discussed below;

    c. Grant permission for entry into the United States of civil SAR facilities of other countries; and

    d. Make arrangements with appropriate customs, immigration, health or other authorities to expedite entry of foreign civil SAR facilities as appropriate.

73. SAR Coordinators will authorize their RCCs to arrange promptly, or in advance, for entry of foreign rescue units into the United States should it ever become necessary. Such arrangements should involve appropriate United States authorities as well as proper authorities of the nation or SAR facility involved with the entry. Such entry may include overflight or landing of SAR aircraft and similar accommodation of surface (land or water) civil SAR units as circumstances dictate. Such entry can be for the purpose of fueling, medical, delivery of survivors, or other appropriate and available operational support, or it could also be in response to a request from a United States RCC to the RCC of another nation for assistance of those facilities.
74. Establishment of jointly sponsored and staffed RCCs or RSCs, is encouraged where appropriate.
75. Use of any SAR facility committed to support any civil SAR mission should normally be coordinated and, as appropriate, directed by an appropriate RCC or RSC consistent with this Plan's provisions.
76. On scene coordination may be delegated to any appropriate unit participating in a particular incident under the cognizance of the SAR mission coordinator or an incident commander.
77. If an RSC is established by any agency, it must operate under the oversight of an RCC, and be responsible for certain tasks, or for portions of the RCC's SRR, as determined by the agency concerned.
78. SAR Coordinators shall arrange for the receipt of distress alerts originating from within SRRs for which they are responsible, and ensure that every RCC and RSC can communicate with persons in distress, with SAR facilities, and with other RCCs/RSCs.

National Incident Management System

79. The type of incident command system adopted for use within the United States is the National Incident Management System (NIMS). When civil SAR operations are conducted in situations where the NIMS structure has been implemented, one or more representatives of the SAR mission coordinator should be assigned to work with the Operations Section of the active incident command post or unified command. Coordination procedures of this Plan will continue to be used under NIMS.

Military Roles and Military/Civilian Relationships

80. Arrangements between federal military and civil agencies should provide for the fullest practicable cooperation, consistent with statutory responsibilities and authorities and assigned civil SAR functions.
81. Cooperative arrangements involving Department of Defense and Coast Guard commands should provide for the fullest practicable use of their facilities for civil SAR on a noninterference basis with primary military duties, consistent with statutory responsibilities and authorities and assigned agency functions.
82. Participants with operational responsibilities should develop plans and procedures for effective use of all available civil SAR facilities and for contingencies to continue civil SAR operations if military forces are withdrawn because of another emergency or a change in military missions.
83. Department of Defense responsibilities under this Plan include support of civil SAR on a non-interference basis with primary military duties, according to applicable national directives, plans, guidelines, agreements, etc.

Resources

84. To optimize delivery of efficient and effective services, and, where practicable and consistent with agency authorities, provide organizations and persons interested in supporting these services the opportunity to do so, all available resources will be used for civil SAR. Certain state and local governments, civil and volunteer organizations, and private enterprises have facilities that contribute to the effectiveness of the over-all civil SAR network, although they are not Participants to this Plan.
85. To help identify, locate and quantify primary civil SAR facilities, Participants may designate facilities which meet international standards for equipment and personnel training as "SAR units" (SRUs). (Such facilities do not need to be dedicated exclusively to the associated type of operations, and this designation is not intended to preclude use of other resources.)

86. Recognizing the critical role of communications in receiving distress information and coordinating a response which can involve multiple organizations and jurisdictions, the Participants will work aggressively to develop suitable civil SAR provisions for:

    a. Interoperability;

    b. Means of sending and receiving alerting information;

    c. Means of identification;

d. Effective provisions for equipment registration and continual access to registration data by civil SAR authorities;

e. Rapid, automatic, and direct routing of emergency communications;

f. High system reliability; and

g. Preemptive or priority processing of distress communications.

Technical and Support Services

87. The Participants will strive together to:

a. Apply the most effective systems to save the most lives at the least operational risk and cost; and

b. Foster innovation in technical, administrative and information systems, which will improve the ability of the Participants and associated non-governmental organizations to carry out civil SAR responsibilities.

88. Each Participant's managerial, operational, and support personnel will strive towards maximizing operational effectiveness.

89. Priority goals of the Participants shall include:

a. Make distress alerts and associated data available to operational personnel as quickly, comprehensively, and reliably as possible;

b. Provide communications systems which are highly reliable, simple, problem-free, interoperable, and as functionally effective as possible;

c. Enable operational personnel to be as highly effective in the planning and conduct of civil SAR operations as possible; and

d. Ensure civil SAR personnel are provided quality training, equipment, procedures, facilities, information, and other tools necessary to conduct SAR operations in a consistent, highly professional, and effective manner.

90. Participants should:

a. Encourage development and proficiency in civil SAR techniques and procedures by other agencies participating in SAR, and assist these other agencies as appropriate;

b. Encourage continued development of State and local civil SAR facilities as appropriate; and

c. Enter into agreements, as appropriate, with State, local, and private organizations to provide for the fullest practicable cooperation in civil SAR consistent with their capabilities and resources, and to account for use of federal facilities in SAR missions with which these organizations are involved.

Suspension or Termination of Operations

91. Civil SAR operations shall normally continue until all reasonable hope of rescuing survivors has passed.
92. The responsible RCC/RSC concerned shall normally decide when to discontinue these operations. If no such center is involved in coordinating the operations, the On Scene Commander (OSC) or Incident Commander (IC) may make this decision. If there is no OSC or IC involved, the decision will be made at an appropriate level of the chain-of-command of the facility conducting the operations.
93. When an RCC/RSC or other appropriate authority considers, on the basis of reliable information that a rescue operation has been successful, or that the emergency no longer exists, it shall terminate the SAR operation and promptly so inform any authority, facility or service which has been activated or notified.
94. If an operation on-scene becomes impracticable and the RCC/RSC or other appropriate authority concludes that survivors might still be alive, it may temporarily suspend the on-scene activities pending further developments, and shall promptly so inform any authority, facility or service which has been activated or notified. Information subsequently received shall be evaluated and operations resumed when justified on the basis of such information.

ENTRY INTO FORCE, AMENDMENT OR TERMINATION

95. Entry into force, amendment or termination of this Plan shall be in accordance with the procedures set forth in the *United States National Search and Rescue Committee Interagency Agreement.*

NATIONAL SEARCH AND RESCUE PLAN OF THE UNITED STATES

Secretary,
Department of Homeland Security

Date: 4-18-07

Secretary,
Department of Commerce

Date:

Deputy Secretary of Defense,
Department of Defense

Date: JUL 18 2007

Secretary,
Department of the Interior

Date: JUN 08 2007

Secretary,
Department of Transportation

Date: 6/15/07

Administrator,
National Aeronautics and Space Administration

Date: AUG 02 2007

Chairman,
Federal Communications Commission

Date: 5/15/07

Appendix J: Emergency Support Function #9 – SAR Annex

ESF Coordinator:
Department of Homeland Security/Federal Emergency Management Agency

Primary Agencies:
Department of Homeland Security/Federal Emergency Management Agency
Department of Homeland Security/U.S. Coast Guard
Department of the Interior/National Park Service
Department of Defense

Support Agencies:
Department of Agriculture
Department of Commerce
Department of Defense
Department of Health and Human Services
Department of Homeland Security
Department of the Interior
Department of Justice
Department of Labor
Department of Transportation
National Aeronautics and Space Administration
U.S. Agency for International Development

INTRODUCTION

Purpose

Emergency Support Function (ESF) #9 – Search and Rescue (SAR) rapidly deploys Federal SAR resources to provide lifesaving assistance to State, tribal, and local authorities, to include local SAR Coordinators and Mission Coordinators, when there is an actual or anticipated request for Federal SAR assistance.

Scope

During incidents or potential incidents requiring a unified SAR response, Federal SAR responsibilities reside with ESF #9 primary agencies that provide timely and specialized SAR capabilities. Support agencies provide specific capabilities or resources that support ESF #9. Federal SAR response operational environments are classified as:

- Structural Collapse (Urban) Search and Rescue (US&R)
- Maritime/Coastal/Waterborne Search and Rescue
- Land Search and Rescue

SAR services include distress monitoring, incident communications, locating distressed personnel, coordination, and execution of rescue operations including extrication and/or evacuation, along with providing medical assistance and civilian services through the use of public and private resources, to assist persons and property in potential or actual distress.

Structural Collapse (Urban) Search and Rescue (US&R)
Primary Agency: Department of Homeland Security (DHS)/Federal Emergency Management Agency (FEMA)
Operational Overview: US&R includes operations for natural and manmade disasters and catastrophic incidents, as well as other structural collapse operations that primarily require DHS/FEMA US&R task force operations. The National US&R Response System integrates DHS/FEMA US&R task forces, Incident Support Teams (ISTs), and technical specialists. The Federal US&R response integrates DHS/FEMA task forces in support of unified SAR operations conducted following the U.S. National Search and Rescue Plan (NSP). (The NSP is the policy guidance of the signatory Federal departments and agencies for coordinating SAR services to meet domestic needs and international commitments.)

DHS/FEMA develops national US&R policy, provides planning guidance and coordination assistance, standardizes task force procedures, evaluates task force operational readiness, funds special equipment and training within available appropriations, and reimburses, as appropriate, task force costs incurred as a result of ESF #9 deployment.

The National US&R Response System is prepared to deploy and initiate operations immediately in support of ESF #9. The task forces are staffed primarily by emergency services personnel who are trained and experienced in collapsed structure SAR operations and possess specialized expertise and equipment. Upon activation under the *National Response Framework (NRF)*, DHS/FEMA US&R task forces are considered Federal assets under the Homeland Security Act of 2002, the Robert T. Stafford Disaster Relief and Emergency Assistance Act, and other applicable authorities.

ISTs provide coordination and logistical support to US&R task forces during emergency operations. They also conduct needs assessments and provide technical advice and assistance to State, tribal, and local government emergency managers.

DHS/FEMA reimburses the parent Sponsoring Agencies for US&R task forces for authorized US&R deployments. DHS/FEMA is authorized to reimburse such activities when there is a Stafford Act declaration or in anticipation of a declaration. For non-Stafford Act US&R deployments, the Federal department or agency requesting US&R assistance reimburses DHS/FEMA following provisions contained in the Financial Management Support Annex. DHS/FEMA uses the funding provided by the requesting Federal department or agency to reimburse the Sponsoring Agency for the task forces.

Maritime/Coastal/Waterborne Search and Rescue

Primary Agency: DHS/U.S. Coast Guard (USCG)

Operational Overview: Maritime/coastal/waterborne SAR includes operations for natural and manmade disasters that primarily require DHS/USCG air, cutter, boat, and response team operations. The Federal maritime/coastal/waterborne SAR response integrates DHS/USCG resources in support of unified SAR operations conducted per the NSP.

DHS/USCG personnel are trained and experienced in maritime/coastal/waterborne SAR operations and possess specialized expertise, facilities, and equipment for conducting an effective response to distress situations. DHS/USCG develops, maintains, and operates rescue facilities for SAR in waters subject to U.S. jurisdiction and is designated the primary agency for maritime/coastal/waterborne SAR under ESF #9. In addition, DHS/USCG staffing at Area, District, and local Sector Command Centers promotes interagency coordination with State, tribal, and local emergency managers during incidents requiring a unified SAR response in which maritime/coastal/waterborne SAR resources allocation are required.

Land Search and Rescue

Primary Agency: Department of the Interior (DOI)/National Park Service (NPS); Department of Defense (DOD)

Operational Overview: Land SAR includes operations that require aviation and ground forces to meet mission objectives, other than maritime/coastal/waterborne and structural collapse SAR operations as described above. Land SAR primary agencies integrate their efforts to provide an array of diverse capabilities under ESF #9.

DOI/NPS possesses SAR resources that are specially trained to operate in various roles including ground search, small boat operations, swiftwater rescue, helo-aquatic rescue, and other technical rescue disciplines. DOI/NPS maintains preconfigured teams that include personnel and equipment from DOI/NPS, U.S. Fish and Wildlife Service, U.S. Geological Survey, Bureau of Indian Affairs, and other DOI components in planning for ESF #9

When requested, DOD, through U.S. Northern Command (USNORTHCOM) and/or U.S. Pacific Command (USPACOM), coordinates facilities, resources, and special capabilities that conduct and support air, land, and maritime SAR operations according to applicable directives, plans, guidelines, and agreements. Per the NSP, the U.S. Air Force and USPACOM provide resources for the organization and coordination of civil SAR services and operations within their assigned SAR regions and, when requested, to assist Federal, State, tribal, and local authorities.

DOD's role as a primary agency is based on SAR Coordinator responsibilities stipulated in the NSP and is generally limited to a coordination function.

DOD designation as a primary agency in ESF #9 is not clearly defined in current statutes, authorities, or DOD policies. Under the *NRF*, DOD assists civil authorities by conducting SAR missions on a reimbursable basis pursuant to the Stafford Act or Economy Act, as appropriate.

If DOD SAR capabilities deploy at the direction of the Air Force Rescue Coordination Center in support of the NSP, and subsequently if the Stafford Act is invoked, those capabilities are administered by the *NRF* and ESF #9. As soon as practical, a DHS/FEMA or other department/agency mission assignment are submitted to and approved by DOD for those capabilities' continued support.

Policies

Federal SAR responders assist and support State, tribal, and local SAR capabilities in incidents requiring a coordinated Federal response. No provision of this annex is to be construed as an obstruction to prompt and effective action by any agency to assist persons in distress.

ESF #9 SAR operations are conducted following the *NRF* and NSP, and the U.S. National SAR Supplement (NSS), Catastrophic Incident SAR (CISAR) Addendum, and other addenda that define SAR responsibilities and provide guidance to the Federal departments and agencies with civil SAR mandates.

If an affected State, tribal, or local government publishes guidance or a plan for conducting unified SAR operations, that guidance or plan takes precedence.

State-to-State SAR assistance is requested by the affected State through the Emergency Management Assistance Compact (EMAC). Other local SAR resources are requested by the affected locality through mutual aid and assistance agreements. Non-Federal SAR resources are, as appropriate, incorporated into any coordinated SAR operations.

State, tribal, and local authorities are responsible for SAR within their respective jurisdictions and typically designate a SAR Coordinator to provide integration and coordination of all SAR services.

The following provides primary agency statutory authorities and policy guidance:

- Homeland Security Act of 2002 (as amended); 6 U.S.C. 722: This section codified US&R as a system within FEMA, "There is in the Agency a system known as the Urban Search and Rescue Response System."
- Stafford Act; 42 U.S.C. 5121-5207: This act authorizes the President (assisted by DHS/FEMA) to declare major disasters and emergencies in the United States and provide assistance to State and local governments. The President may use the services of State and local governments for the

purposes of the act, which includes addressing immediate threats to life and property (e.g., SAR operations).

- 14 U.S.C. 2: This section requires the U.S. Coast Guard to develop, establish, maintain, and operate rescue facilities for the promotion of safety on, under, and over the high seas and waters subject to the jurisdiction of the United States.
- 16 U.S.C. 1b(1): This section gives DOI/NPS authority to provide emergency rescue, firefighting, and cooperative assistance to public safety agencies for related purposes outside of the National Park System.
- Economy Act; 31 U.S.C. 1535-1536 (2007): This act authorizes Federal departments and agencies to provide goods or services, on a reimbursable basis, to other Federal departments and agencies.
- 32 U.S.C.: This title authorizes the National Guard to perform DOD-funded activities while remaining under the control of the Governor.
- Post-Katrina Emergency Management Reform Act; P.L. 109-295 (2006): This act expands the scope of ESF #9 from only urban SAR to include all types of SAR activities. Follow on congressional guidance establishes the organizational structure. It codified US&R as a system within FEMA in the Homeland Security Act of 2002 (as amended). It also mandated FEMA to develop a Federal response capability to rapidly and effectively deliver assistance essential to saving lives or protecting property or public health and safety and to carry out the mission of FEMA by conducting emergency operations to save lives and property.
- National Search and Rescue Plan (NSP): The NSP is the policy guidance of the signatory Federal departments and agencies for coordinating SAR services to meet domestic needs and international commitments.
- National SAR Supplement (NSS): This document provides implementation guidance on the International Aeronautical and Maritime Search and Rescue Manual and the NSP.
- Catastrophic Incident SAR (CISAR) Addendum to the NSS: This document provides a description of the unified SAR response to catastrophic incidents, guides Federal authorities involved in the response, and informs State, tribal, and local authorities on what to expect of/from Federal SAR responders.
- DOD Support to Civil Search and Rescue (DODD 3003.01): This directive states that DOD shall support domestic civil authorities by providing civil SAR service to the fullest extent practicable on a noninterference basis with primary military duties.
- Military Support to Civil Authorities (DODD 3025.1): This directive identifies the policy and responsibilities by which DOD responds to major disasters or emergencies per the Stafford Act and other authorities.
- Military Assistance to Civil Authorities (DODD 3025.15): This directive states that DOD shall cooperate with and provide military assistance to civil authorities, as directed by and consistent with applicable law, Presidential directives, and Executive orders.

CONCEPT OF OPERATIONS

General

DHS/FEMA activates ESF #9 when an incident is anticipated or occurs that may result in a request for a unified SAR response to an affected area. The ESF #9 response is scalable to meet the specific needs of each incident, based upon the nature and magnitude of the event, the suddenness of onset, and the capability of local SAR resources. Response resources are drawn from ESF #9 primary and support agencies.

As required, the primary agencies are represented at the National Response Coordination Center (NRCC), Joint Field Office (JFO), and State, tribal, and local Emergency Operations Centers (EOCs).

For each incident requiring Federal SAR support, DHS/FEMA designates the overall primary agency for that particular ESF #9 SAR response. Designation is dependent upon incident circumstances and the type of response required.

The designated overall primary agency coordinates integration of Federal SAR resources, including support agency resources, in support of the requesting Federal, State, tribal, or local SAR authority.

All ESF #9 agencies provide support to the designated overall primary agency, as required.

ORGANIZATION

For incidents where DHS/FEMA is the overall primary agency, ESF #9 SAR operations are conducted following the National US&R Response System manuals, directives, NSP, NSS, and CISAR Addendum.

For incidents where DHS/USCG is the overall primary agency, ESF #9 SAR operations are conducted following the SAR response structure as outlined in the NSP, NSS, CISAR Addendum, USCG SAR Addendum, and other USCG directives.

For incidents where DOI/NPS and/or DOD are the overall primary agency, ESF #9 SAR operations are conducted following the SAR response structure as outlined in the NSP, NSS, CISAR Addendum, and other relevant DOI/NPS and DOD SAR procedures, directives, and manuals.

RESPONSIBILITIES

ESF #9 Coordinator: DHS/FEMA

As ESF #9 coordinator, DHS/FEMA:

- Designates the overall primary agency responsible for the coordination of Federal SAR operations.
- Coordinates with all other ESFs, as required.

Primary Agencies

For every incident, DHS/FEMA assesses the specific SAR requirements and assigns one of the four primary agencies as the overall primary agency for SAR for that particular incident.

When in the overall primary agency role for a particular incident, that organization conducts the following actions:

- Coordinates planning and operations between primary and support agencies.
- Coordinates resolution of conflicting operational demands for SAR response resources.

Primary Agency: DHS/FEMA

- DHS/FEMA serves as the overall primary agency to accomplish the ESF #9 mission during structural collapse SAR operations in incidents requiring a coordinated Federal response.
- For incidents in which it is designated the overall primary agency, DHS/FEMA: Manages US&R task force and IST deployments in the affected area.
- Coordinates logistical support for US&R assets during field operations.
- Coordinates the provisioning of additional support assets.
- Coordinates with Federal, State, tribal, and local designated SAR authorities to integrate Federal SAR resources.
- As required, provides representation at the NRCC, JFO, and State, tribal, and local EOCs.
- Provides incident reports, assessments, and situation reports as required.

Primary Agency: DHS/USCG

DHS/USCG serves as the overall primary agency to accomplish the ESF #9 mission during maritime/coastal/waterborne SAR operations in incidents requiring a coordinated Federal response.

For incidents in which it is designated the overall primary agency, DHS/USCG:

- Manages USCG SAR resources in the affected area.
- Coordinates the provisioning of additional support assets.
- Coordinates with Federal, State, tribal, and local designated SAR authorities to integrate Federal SAR resources.
- As required, provides representation at the NRCC, JFO, and State, tribal, and local EOCs.
- Provides incident reports, assessments, and situation reports, as required.

Primary Agency: DOI/NPS

DOI/NPS and DOD share responsibility as the overall primary agency for a particular incident to accomplish the ESF #9 mission during land SAR operations in incidents requiring a coordinated Federal response.

For incidents in which it is designated the overall primary agency, DOI/NPS:

- Manages DOI/NPS land SAR resources in the affected area.
- Coordinates the provisioning of additional support assets.
- Coordinates with Federal, State, tribal, and local designated SAR authorities to integrate Federal SAR resources.
- Coordinates logistical support for DOI/NPS resources during field operations.
- As required, provides representation at the NRCC, JFO, and State, tribal, and local EOCs.
- Provides incident reports, assessments, and situation reports as required.

Primary Agency: DOD

DOD and DOI/NPS share responsibility as the overall primary agency for accomplishing the ESF #9 mission during land SAR operations in incidents requiring a coordinated Federal response.

For incidents in which it is designated the overall primary agency, DOD, through USNORTHCOM and USPACOM:

- Manages DOD SAR resources in the affected area.
- Coordinates the provisioning of additional support assets.
- Coordinates with Federal, State, tribal, and local designated SAR authorities to integrate Federal SAR resources.
- As required, provides representation at the NRCC, JFO, and State, tribal, and local EOCs.
- Provides incident reports, assessments, and situation reports as required.

SUPPORT AGENCIES

Agency	Functions
Department of Agriculture	**Forest Service** • Develops standby agreements to provide equipment and supplies from the National Interagency Fire Center (NIFC) Cache System at the time of deployment. • Develops contingency plans for use of NIFC contract aircraft during incidents. • If available, provides equipment and supplies from the NIFC Cache System and use of NIFC contract aircraft.
Department of Commerce	**National Oceanic and Atmospheric Administration** • Acquires and disseminates weather data, forecasts, and emergency information. • Provides weather information essential for efficient SAR. • Predicts pollutant movement and dispersion over time (marine and atmospheric). • Assesses areas of greatest hazard following a marine or atmospheric release. • Provides satellite services for detecting and locating persons in potential or actual distress in the wilderness, maritime, and aeronautical environments.

Agency	Functions
Department of Defense	**National Geospatial-Intelligence Agency (NGA)** • Coordinates and manages the timely tasking, acquisition, analysis, and delivery of satellite imagery or imagery-derived products as directed by the primary agency. • Provides expert analysis of imagery to determine damage levels and other elements of essential information, as needed. • Provides technical expertise/analysis from other imagery sources, if such expertise resides within DOD/NGA. • Provides mobile geospatial intelligence including technical experts (imagery analysts and geospatial analysts) and robust communications to support SAR field teams or other DHS/FEMA field teams, as requested by the primary agency. • Provides imagery-derived and geospatial intelligence analysis in preparation for potential disasters or emergencies. • Coordinates the release and dissemination of DOD/NGA products and/or data following applicable security classifications, licensing, copyright agreements, and limited distribution restrictions.
	U.S. Army Corps of Engineers (USACE) • Under the NRF, supports the ESF #9 SAR mission by developing, training, and equipping USACE personnel to operate as support to the DHS/FEMA US&R Task Forces. • Through Technical Assistance Structural Engineers (TASEs), supports DHS/FEMA and other agency efforts requiring structural engineering expertise (e.g., evaluate, design, construct, or repair of buildings, bridges, and critical facilities). • Through Structural Safety Assessment Planning and Response Teams (SSA PRTs), provides habitability inspections as required, to support response and recovery efforts for building safety evaluations.
Department of Health and Human Services	**National Disaster Medical System (NDMS)** • Through ESF #8 – Public Health and Medical Services, provides support to ESF #9 primary agencies, including liaisons; medical supplies, equipment, and pharmaceuticals; supporting personnel; and veterinary support. • Provides NDMS personnel to support medical field operations and evacuation. **Indian Health Service**: Maintains specialized response teams to support the medical care of American Indian and Alaska Native people.
Department of Homeland Security	**Customs and Border Protection** • Maintains Border Patrol Search, Trauma, and Rescue (BORSTAR) teams, which are highly specialized units capable of responding to emergency SAR situations anywhere in the United States. • Maintains air and marine assets to support SAR transportation operations.
Department of the Interior	**U.S. Geological Survey:** Provides personnel with appropriate technical disciplines and specialized technology to support geospatial analysis and mapping products in support of ESF #9 primary agencies.

Department of Justice	• Coordinates force protection as required. • Provides assistance with the development and maintenance of tort liability claims coverage for US&R task force and IST personnel engaged in mobilization, deployment, and field operations.
Agency	**Functions**
Department of Labor	• The Mine Safety and Health Administration provides mine rescue teams, mobile command centers, seismic location systems, TV probe systems, gas sampling analysis, and robot explorers. • The Department of Labor Employment Standards Administration, through its Federal Employees' Compensation Program, provides workers compensation guidance, claims resolution, and coverage for US&R task force and IST personnel while they are engaged in mobilization, deployment, and field operations. • The Occupational Safety and Health Administration implements procedures contained in the Worker Safety and Health Support Annex to provide onsite technical assistance, including the evaluation of SAR team exposure to hazardous substances and the dangers of structural collapse.
Department of Transportation	**Federal Aviation Administration** • Is delegated sole authority to manage the National Airspace System (NAS), which includes operating a safe, secure, and efficient air traffic system; oversight and certification of aircraft and airmen; regulation of airspace; promotion of air commerce, and the support of America's national defense (49 U.S.C.). • Supports activities to protect and recover NAS operations.
National Aeronautics and Space Administration	• Provides personnel in appropriate technical disciplines (e.g., its Disaster Assistance and Rescue Team). • Provides temporary use of facilities for mobilization centers and staging areas for SAR assets.
U.S. Agency for International Development	Manages the support of international SAR teams to a domestic U.S. disaster following a Stafford Act Declaration, under the International Assistance System Concept of Operations (IAS CONOPS), and in support of the NRF's International Coordination Support Annex (ICSA.)
Department of State	If FEMA does not active the IAS and proactive offers of assistance from foreign countries or international/multilateral organizations are received, the State Department may still designate the State Task Force (STF) (or, if the STF has not been established, designate a lead bureau or the Operations Center's Crisis Management Support [CMS] office) as the sole entity within DOS responsible for coordinating formal offers of international assistance, and request all offers be forwarded to the STF (or, as appropriate, the lead bureau or CMS office) for dispensation.

This page intentionally left blank.

Index

D

E

F

G

H

I

O

P

R

S

T

U

V

W

Other Selected Books from dbS Productions

Land Search and Rescue Manual

Volume I: Land Search and Rescue Addendum by NSARC
ISBN 978-1-879471-43-6 2012 304p 8.5 x 11 paperback $25.00
e-book ISBN 978-1-879471-44-3 $9.99

Volume II: Catastrophic Incident Search and Rescue Addendum by NSARC
ISBN 978-1-879471-45-0 2012 262p 8.5 x 11 paperback $25.00
e-book ISBN 978-1-879471-46-7 $9.99

Volume III: Incident Command System Field Operations Guide for Search and Rescue by Robert J. Koester
ISBN 978-1-879471-47-4 2012 216p 4.25 x 7 wiro $25.00
e-book ISBN 978-1-879471-48-1 $9.99

Volume IV: Management Field Operations Guide for Search and Rescue
by Robert J. Koester
ISBN 978-1-879471-49-8 2012 xxxp 4.25 x 7 paperback $25.00
e-book ISBN 978-1-879471-50-4 $9.99

Volume V: Tactical Field Operations Guide for Search and Rescue
by Robert J. Koester
ISBN 978-1-879471-51-1 2012 xxxp 4.25 x 7 paperback $25.00
e-book ISBN 978-879471-52-8 $9.99

Lost Person Behavior: A Search and Rescue Guide on Where to Look – for Land, Air and Water by Robert J. Koester
ISBN 978-1-879471-39-9 2008 416p 5.5 x 8.5 paperback with semi-concealed wiro $25.00

The Handbook for Aviation Survival Sense by Robert "Skip" Stoffel and Brett C. Stoffel
2009 160p 4.25 x 7 waterproof paper wiro $28.00

Urban Search: Managing Missing Person Searches in the Urban Environment
by Christopher S. Young and John Wehbring
ISBN 978-1-879471-38-2 2007 352p 5.5 x 8.5 paperback $25.00

Foundations for Awareness, Signcutting and Tracking by Robert Speiden
ISBN: 978-0-9817686-0-1 2009 268p 8.5 x 11 paperback $35.00

Incident Commander for Ground Search and Rescue: A Training Course for Incident Commanders by Robert J. Koester
Student Manual ISBN 978-1-879471-22-1 1997 230p 8.5 x 11 $35.00
Instructor Manual ISBN 978-1-879471-21-4 1999 290p 8.5 x 11 $75.00

Made in the USA
Coppell, TX
16 November 2025

63306102R00171